Water Lords

China Studies

PUBLISHED FOR THE INSTITUTE FOR CHINESE STUDIES, UNIVERSITY OF OXFORD

Series Editor

Rana Mitter, *Harvard Kennedy School*

VOLUME 54

The titles published in this series are listed at *brill.com/chs*

Water Lords

Beijing's Drinking Water and Its Carriers, 1644–1937

Lei Zhang

BRILL

LEIDEN | BOSTON

Cover illustration: Beijing Water Cart, circa 1917–1919. Sydney D. Gamble Photographs Collection, Duke University Library.

Library of Congress Cataloging-in-Publication Data

Names: Zhang, Lei (Writer on historical geography), author.
Title: Water lords : Beijing's drinking water and its carriers, 1644-1937 / Lei Zhang.
Description: Leiden ; Boston : Brill, [2025] | Series: China studies, 1570-1344 ; volume 54 | Includes bibliographical references and index. |
Identifiers: LCCN 2024046008 (print) | LCCN 2024046009 (ebook) | ISBN 9789004693241 (hardback) | ISBN 9789004720800 (ebook)
Subjects: LCSH: Water-supply–China–Beijing–History–20th century. | Water carriers (Persons)–China–Beijing–History–20th century. | Drinking water–China–Beijing–History–20th century. | City and town life–China–Beijing–History–20th century.
Classification: LCC TD302.B45 Z434 2025 (print) | LCC TD302.B45 (ebook) | DDC 363.6/109511560904--dc23/eng/20241219
LC record available at https://lccn.loc.gov/2024046008
LC ebook record available at https://lccn.loc.gov/2024046009

Typeface for the Latin, Greek, and Cyrillic scripts: "Brill". See and download: brill.com/brill-typeface.

ISSN 1570-1344
ISBN 978-90-04-69324-1 (hardback)
ISBN 978-90-04-72080-0 (e-book)
DOI 10.1163/9789004720800

Contents

Figures and Tables

Figures

Tables

Introduction

In Late Imperial China, Westerners traveling to Beijing were impressed with its massive number of wells, as evidenced in Scottish medical missionary John Dudgeon's 1870s description:

> In respect to wells, Peking is probably unsurpassed by any city in the world. They exist in almost every lane; and in the great thoroughfares, four, five or more may be met within a mile. They are never exhausted. They are usually situated on alternate sides of the streets, and are for the most part surrounded by willows. Almost every house of any consequence also contains one or two wells.[1]

No city has been able to exist without subjugating water to its demands. But unlike the cities of south China whose drinking water was mainly procured from rivers, early modern Beijing relied exclusively on wells to extract groundwater. If the city of Rome owed its existence to its impressive aqueducts, Beijing was indeed the city of wells.

As the city's essential water infrastructure, hand-dug wells of thirty feet in depth provided an abundant supply of potable water and made Beijing possible. Unfortunately, Beijing was also historically notorious for the saltiness and turbidity of its drinking water. The water carried a great deal of lime carbonates, magnesium, and other dissolved minerals that dramatically contributed to its hardness and brackishness. What was worse, these shallow wells were open to surface pollutants such as percolation from latrines. Because the water drawn from many of the city's wells was too poor in quality to quench the city's thirst, it was used for other purposes or mixed with higher quality water to make it drinkable. The predominance of unpalatable water in Beijing thus required a distribution system to deliver drinkable water, and this necessitated an elaborate supply and distribution network that was more intricately constructed than historians have recognized. Moreover, unlike the city of London before 1580, where citizens freely drew the water they needed, drinking water had always been a costly and contested commodity in densely populated Beijing, which was located on the semiarid north China plain. Thus, professional water carriers became necessary. These water carriers transported water from the wells to the households of city dwellers according to a sophisticated delivery system developed in the early Qing dynasty (1644–1911).

1 C. A. Gordon, *An Epitome of the Reports of the Medical Offices to the Chinese Imperial Maritime Customs Service, from 1871 to 1882* (London, 1885), 3.

 | DOI:10.1163/9789004720800_002

Drinking water has become an increasingly pressing issue in cities all over the world and historians have conducted extensive research on this subject for decades.[2] Drinking water is not simply a physical process, but one imbued with political, economic, and social meanings. Nevertheless, the conceptual and empirical understanding of urban drinking water, which is mainly grounded in the European and North American experience, assumes that the modernization of a city's water occurs along with the establishment of modern urban governance. This has created a discontinuity between premodern and modern water history in which water systems and water vendors in premodern cities are considered rudimentary and have thus slipped from view. But for cities with different relations to tradition and imperialism, the modernization of city water did not necessarily follow the pattern prevalent in Europe and North America. A prominent example is Istanbul, where the Ottomans built public fountains to ensure a secure supply of water; these fountains coexisted with modern centralized waterworks for six decades, making this dual system a characteristic feature of Istanbul's modern water supply.[3]

In the Chinese context, water is likewise a central theme in hydraulic history, as captured by Karl A. Wittfogel's term "hydraulic empire," which ascribed to water a decisive role in determining political forms of despotic government.[4] Joseph Needham also contended that hydraulic works were the technological heart of China's civilization, and before the nineteenth century no country could match China's ability to control its surface water.[5] To a large extent, the foundation of the Chinese empire was its control of water, which

2 Much research has been done on drinking water in cities. See, for example, Nelson M. Blake, *Water for the Cities: A History of the Urban Water Supply Problem in the United States* (Syracuse: Syracuse University Press, 1956); Asok Kumar Mukhopadhyay, *Politics of Water Supply: The Case of Victorian London* (Calcutta: World Press, 1981); Jean-Pierre Goubert, *The Conquest of Water: The Advent of Health in the Industrial Age* (Princeton, NJ: Princeton University Press, 1989); Theodore Steinberg, *Nature Incorporated: Industrialization and the Waters of New England* (Amherst, MA: University of Massachusetts Press, 1991); Vivienne Bennett, *The Politics of Water: Urban Protest, Gender and Power in Monterrey, Mexico* (Pittsburgh, PA: University of Pittsburgh Press, 1995); Gerard T. Koeppel, *Water for Gotham: A History* (Princeton, NJ: Princeton University Press, 2000); Carl Smith, *City Water, City Life: Water and the Infrastructure of Ideas in Urbanizing Philadelphia, Boston, and Chicago* (Chicago: University of Chicago Press, 2013); John Broich, *London: Water and the Making of the Modern City* (Pittsburgh, PA: University of Pittsburgh Press, 2013).

3 Noyan Dinçkal, "Reluctant Modernization: The Cultural Dynamics of Water Supply in Istanbul, 1885–1950," *Technology and Culture* 49, no. 3 (July 2008): 675–700.

4 Karl A. Wittfogel, *Oriental Despotism: A Comparative Study of Total Power* (New Haven and London: Yale University Press, 1967).

5 Joseph Needham, *Science and Civilization in China, vol. 4* (Cambridge: Cambridge University Press, 1954).

requires central coordination and a specialized bureaucracy. In particular, the management of the turbulent Yellow River for flood control and irrigation on a gigantic scale has been seen as a major indicator of state performance and a barometer of political legitimacy in China.[6] However, the existing scholarship on water governance in historical China leaves two major lacunae waiting to be filled. First, it focuses on large-scale hydraulic projects and neglects the smaller-scale task of ensuring regular and predictable access to drinking water in urban China. Second, it overlooks the role of water as a daily necessity; a dependable and plentiful supply of pure water for a large population is vital to sustaining life.[7] As a consequence, the significance of potable water to city life in China remains an unrecognized and unappreciated topic in English-language literatures.[8]

Historians frequently discuss the issue of state power in Beijing because, as an imperial and republican capital, it was the center of a political network that

---

6 To name but a few, Pierre-Etienne Will, "State Intervention in the Administration of a Hydraulic Infrastructure: The Example of Hubei province in Late Imperial Times," in Stuart Schram, ed., *The Scope of State Power in China* (London: SOAS, 1985); Randall Dodgen, *Controlling the Dragon: Confucian Engineers and the Yellow River in Late Imperial China* (Honolulu: University of Hawai'i Press, 2001); David A. Pietz, *The Yellow River: The Problem of Water in Modern China* (Cambridge, MA: Harvard University Press, 2015); Micah S. Muscolino, *The Ecology of War in China: Henan Province, the Yellow River and Beyond, 1938–1950* (Cambridge: Cambridge University Press, 2015). Nevertheless, the theory of China as a hydraulic empire has been critiqued by historians who have conceptualized the Chinese water management and water rights in local society, especially in terms of water supplies, using "hydraulic society" and "hydraulic cycle" models. These historians underline the power of local autonomous communities, particularly in rural China, rather than the power of the imperial bureaucratic state. See Pierre-Etienne Will, "Clear Waters versus Muddy Waters: The Zheng-Bai Irrigation System of Shaanxi Province in the Late Imperial Period," in Mark Elvin and Ts'ui-jun Liu, eds., *Sediments of Time: Environment and Society in Chinese History* (Cambridge: Cambridge University Press, 1998) 283–343; Peter Perdue, *Exhausting the Earth: State and Peasant in Hunan, 1500–1850* (Cambridge, MA: Harvard University Press, 1987); Keith Schoppa, *Xiang Lake: Nine Centuries of Chinese Life* (New Haven, CT: Yale University Press, 1989).

7 In the history of food, there have been many works on sustaining the body with food, but fewer on sustaining it with water. This book addresses water control history, with water as a consumable good.

8 Regarding the history of drinking water in Beijing, Chung-lin Chiu outlined the history of the water supply and its carriers from late Ming to Republican periods while Yuanbao Xiong studied the water distribution territories in the Qing period; see C. L. Chiu, "Shuiwozi: Beijing de minsheng yongshui, 1637–1937," in H. T. Li, ed., *Zhongguo chengshi shenghuo,* 203–52 (Beijing: Xinxing chubanshe, 2006); Yuanbao Xiong, "Shindai Pekin no suidoji," in Yuanbao Xiong ed., *Shindai Huishu chiiki shakkaishi kenkyu,* 291–327 (Tokyo: Kyuko Shoin Press, 2003).The other related research includes Mingzheng Shi, "Beijing Transformations: Urban Infrastructure, Public Works, and Social Change in the Chinese Capital, 1900–1928" (Ph.D. diss., Columbia University, 1993).

connected it to the rest of the Chinese provinces. However, there was another form of power in the city: the "water power" of the water carriers who distributed the city's drinking water to the residents of Beijing. This unconventional avenue of power was nonetheless successfully exercised by water carriers in the alleyways beyond the Forbidden City in the Qing period. Although neglected by mainstream history, water carriers constructed a hidden water regime in the neighborhoods across the city and were major actors in the daily life of Beijing. After 1900, they demonstrated versatility by co-opting new well-drilling technology, manipulating political movements, and reorganizing their well guild to better solidify their power in order to deal with other actors in Beijing, including the city's inhabitants, a newly established running water company called the Capital Waterworks (京師自來水公司), and the state. In the 1920s, Beijing residents crafted the term "water lord" (水閥) from the term "warlord," which refers to the warlord era of China between 1916 and 1928. Warlords were persons who divided regions of China among themselves and espoused the ideal that militancy was necessary to legitimize their power. The water lords of Beijing controlled water distribution in specific areas of the city and resorted to violence to maintain their control. Thus, the city of Beijing was a place not only for warlords, but also for water lords until the arrival of the Nationalist government in 1928 and its aggressive push to tighten its grip on water and its suppliers through a carefully coordinated sanitation campaign.

This book presents the untold story of drinking water and its carriers in the face of change from the imperial Qing to the Republican period, especially from 1900 to 1937, when the city of Beijing underwent unprecedented modernization and urbanization. Since the 1970s, new studies on urban China have rapidly advanced as scholars have decisively abandoned Max Weber's proposition that Chinese cities mainly served as the residences of imperial viceroys. Urban historians have provided detailed accounts of ordinary people's lives and have suggested that daily life was far richer and more complicated than was previously thought.[9] More recent studies on modern Beijing have begun to get at the experiences and activities of ordinary people beyond official Beijing.[10]

9 William T. Rowe, *Hankow: Commerce and Society in a Chinese City, 1796–1889* (Stanford, CA: Stanford University Press, 1984); William T. Rowe, *Hankow: Conflict and Community in a Chinese City, 1796–1895* (Stanford, CA: Stanford University Press, 1989).

10 Susan Naquin, *Peking: Temples and Life, 1400–1900* (Berkeley: University of California Press, 2000); Yue Dong, *Republican Beijing: The City and Its Histories* (Berkeley: University of California Press, 2003); Vincent Goossaert, *The Taoists of Peking, 1800–1949: A Social History of Urban Clerics* (Cambridge, MA: Harvard University Asia Center, 2007); Janet Y. Chen, *Guilty of Indigence: The Urban Poor in China, 1900–1953* (Princeton, NJ: Princeton University Press, 2012); Zhao Ma, *Runaway Wives, Urban Crimes, and*

However, the city and nature are intertwined, and urban history without its environmental dimension is incomplete since nature penetrates human society and affects everyday life. While water plays an essential role in the modernization of cities like Beijing, historical accounts have traditionally tended to focus on the public health benefits that accompanied the establishment of municipal water supplies; but this substantially downplays the broader impact of the modernization of water. Water networks did more than improve health and showcase the impact of hygienic modernity; they also shaped the political landscape within the city.

This book introduces an alternative account of the history of Beijing by attending to water as a daily necessity and examining the previously neglected arrangements that provided urban denizens with regular and predictable access to this critical resource. Furthermore, this book distinguishes previous historical research on Beijing by foregrounding water in the city setting as a distinctive force shaping urban life in the realms of technology, society, and politics, and examines water governance—including water technology innovations, water crises, and water competition in the city of Beijing—not only from an urban history perspective, but more importantly, from an ecological history perspective.

To conceptualize water governance in historical Beijing, this book uses the concept of the "water regime" as its analytical framework. Inspired by the fire regime,[11] I see the water regime as an environmental infrastructure characterized by a system of hierarchy and control based on water, in which customary law and violence are crucial means of enforcement for the maintenance of that control. The water regime can be utilized to describe and analyze the set of regulatory processes, mechanisms, and organizations through which actors influence environmental actions and outcomes. The water regime explores the materiality of water and its significance in everyday life both in the social and natural realms. It includes not only the actions of the state but also other

---

Survival Tactics in Wartime Beijing, 1937–1949 (Cambridge, MA: Harvard University Press, 2015).

11 The historians of urban conflagrations have used of the term "regime" to describe how the configuration of material and political relations in cities helps to create a fire regime. The fire regime is meant as the nexus of environmental conditions, including climate, topography, and natural resources, with the political system that organizes and sustains concentrated settlement. See Bankoff et al., *Flammable Cities: Urban Conflagration and the Making of the Modern World* (Madison: University of Wisconsin Press, 2012), 8–9. Yan Gao also employed the term "water regime" and broadly defined it as being concerned with historical water bodies and various natural elements in the studies of hydraulic history in the central Yangtze River valley. Yan Gao, *Yangzi Waters: Transforming the Water Regime of the Jianghan Plain in Late Imperial China* (Brill, 2022).

actors, such as communities, that are subsumed within policy networks, multilevel processes, and relations.[12] Unlike the state's unshakable commitment to water control as exemplified in Late Imperial China,[13] the state in modern China adopted a laissez-faire attitude toward water management in the city of Beijing by establishing informal governance to ensure a steady water supply. Rather than fully involving itself, the state placed the water supply in the hands of professional water carriers. This intentional hydraulic negligence catalyzed the rise of the water regime, in which there is a notable shift from the dominant state-centric focus to the role of non-state actors in the underlying arrangements through which society-environment relations are governed.

The water regime as a distinctive ecosystem was sustained by well house systems that were usually composed of wells, well owners with water distribution territories, and water carriers. Water's central role in daily life and its environmental limitations in respect of quality, scarcity, and accessibility allowed the water carriers to create a monopoly and prevent the use of alternative resources. Water carriers manipulated to subjugate residents and the state so that they could control drinking water within their regime. To capture especially important aspects of how the water and its carriers exerted their power to shape modernity's impact in Beijing, this book emphasizes the theme of everyday technology within the framework of the water regime.

Urban life is characterized by a more complex web of interrelations with technology. The fact of the water regime's reliance on underground water reveals significant gaps in the history of water-related technology. For instance, well boring did not simply consist of sinking a hole in the ground; rather, it was a systematic project that encompassed determining a site, boring testing, sinking and casing the well, and regular maintenance of the finished well.[14]

---

12 Maria Carmen Lemos and Arun Agrawal, "Environmental Governance," *Annual Review of Environmental Resources* 31 (2006): 297–325. Mathew Himley, "Geographies of Environmental Governance: The Nexus of Nature and Neoliberalism," *Geography Compass* 2 (2008): 433–451.

13 Ling Zhang, *The River, the Plain, and the State: An Environmental Drama in Northern Song China, 1048–1128* (Cambridge: Cambridge University Press, 2016), 13–14.

14 The history of well-drilling technology in China has not been thoroughly investigated. The existing research tends to highlight the significance of modern waterworks and their technology to the urban water supply, while little attention has been paid to water-well drilling, which was considered outdated and unhygienic. Well-drilling technology has occasionally been addressed in studies of the salt industry in China, see Chunyuan Liu, "Woguo songdai jingyan zuantan." For the most recent research, see Jixing Pan, "Zhongguo shenjing zuantan." For example, Needham's volumes on *Science and Civilization in China* do not devote even one page to well drilling. Needham briefly mentions that in the early Han period (206 BCE–9 CE) the Chinese had fully developed the technique of

More critically, the water regime is not only about technological innovation, but about the persistence of technology, including situations of repair, renovation, and reuse to deliver operability, since infrastructure facilities are in constant need of maintenance to restore material order.[15] Although mainly conducted behind the scenes, maintenance conserves the existing infrastructure, meanwhile it serves to alter the original structure by blending old and new.[16] The water regime represented the accommodation made between the state, technology, and society in both a temporal and a spatial sense, and suggests a more complex relationship between the state and water than mainstream accounts provide.

The theme of everyday technology offers an alternative mode of modernity between the visible and invisible dimensions of urban space.[17] Owing to the uneven accessibility and distribution of water, spatial differentiation was a crucial factor in sustaining the power of the water carriers. My research illustrates how water and its carriers created and imposed a spatial order within their water regime, over which they held near-total dominion. This spatial order was ever-changing under the influence of new technology, social unrest, and state-making; and water carriers constantly needed to define and defend their territories in the daily life of Beijing. Thus, the framework of the water regime conveys a vision of water along with its daily sustainability and availability, and carriers as participants in a dynamic system in which water carriers were constantly renegotiating its role. This not only challenges the linear narrative of modernity in China but also reveals the changeability of modernity.

Through an in-depth investigation of water carriers' lived experiences in a competitive world and an examination of their strategies for coping with risks and opportunities in their everyday encounters with environmental realities, this book argues that the water regime is a social and environmental infrastructure in which the water carriers accommodated social processes, material forces, and negotiations between their own interests and those of the city and the state. The transformation of the water carriers into water lords was prompted by state building, the rise of the modern water enterprise, and the

---

drilling boreholes two thousand feet deep; see Joseph Needham, *Science and Civilization in China*. However, this argument is incorrect. The application of deep drilling in wells occurred around 1041 CE of the Song Dynasty instead of in the Han Dynasty.

15 Heike Weber and Stefan Krebs, eds., *The Persistence of Technology: From Maintenance and Repair to Reuse and Disposal* (transcript Verlag, 2021), 9–25.

16 Stephen Graham and Nigel Thrift, "Out of Order: Understanding Repair and Maintenance," *Theory, Culture & Society* 24, no. 3 (2007): 1–25.

17 Madeleine Yue Dong and Joshua L. Goldstein, eds., *Everyday Modernity in China* (Seattle: University of Washington Press, 2006).

changing nature of Beijing's neighborhoods. In this sense, the book offers an illuminating account of water governance in Beijing and provides a unique window through which to understand, from a bottom-up perspective, the water carriers' roles both as water lords and as proletarians in the formation and transformation of Beijing's urban landscape. This account not only advances our understanding of the resilience of everyday power in urban China, but also sheds light on the dynamic environmental history of Chinese cities during China's transition from empire to republic.

1 The Structure of This Book

First, to frame my inquiry, the first chapter of this book begins with the formation and evolution of the water regime in the Qing period when water carriers established their ecological regime and incorporated city authorities, urban dwellers, and water into their domain. Because the availability and quality of water do not necessarily follow the cycle of political regimes, this book is not organized according to conventional political periodizations, namely the boundaries between the Qing and Republican periods from 1644 to 1937. Yet, due to the availability of historical sources, only one chapter addresses the long period in the Qing dynasty between 1644 and 1900, while the rest of the book examines the short period from 1900 to 1937, during which Beijing underwent unprecedented modernization and was abruptly interrupted by the Japanese occupation.

I delve into the city's relationship to water by looking at water carriers in the neighborhoods where they worked and by exploring who the water carriers were and what constituted the water trade in Beijing. Then, I look at the networks of the water carriers who flooded into Beijing as migrants, and the role of native-place sentiment in the negotiation of power. Finally, I examine the hierarchy of the water carriers' power, including the relations between well owners and water carriers, between water carriers living in different parts of Beijing, and between water carriers from different regions in China. Moreover, I analyze how water carriers shaped Beijing residents' habits of consuming water, and how water carriers' delivery routines were woven into the fabric of the residents' daily lives. After tracing the contours of the water regime in the Qing period, I examine three different contexts of water modernization in which changes in the water source and its governance shaped water and its suppliers: (1) foreign well water, the arrival of Japanese well-drilling technology in the first decade of the twentieth century; (2) tap water, the challenge of competing with the arrival of running water in 1910; and (3) hygienic water, the imposition of state hygiene mandates that began in 1928.

The first context I focus on in chapter two is foreign well water, specifically how the introduction of well-drilling technology that accompanied the processes of modernizing and state-building impacted the water regime in terms of the water supply and its carriers during the last decade of the Qing dynasty (1900–1910). In 1900, when Japanese well drillers brought their well-drilling technology to Beijing, rather than opposing this new technology, water carriers embraced it and became the chief proponents of boring foreign wells as new water sources. In doing so, water carriers successfully internalized the external challenge imposed by Japanese well drilling and maintained their control over the water supply. At the same time, the deep dichotomy between the Inner City and Outer City, which had been segregated by ethnicity since the 1640s, shaped the water carriers' adaptations to new well-drilling technology. More importantly, water carriers would use foreign wells to cope with a new challenge that was just around the corner—competition from running water.

The second context I address in chapters three and four is tap water and the interactions during the Beiyang Government era (1912–1927)[18] between water carriers and their rival, the Capital Waterworks, which was expected to bring tap water to most households in the city and thus had a seemingly strong competitive advantage over wells water and its carriers. Ironically, Beijing's newly created Waterworks relied on water carriers to sell its product in the 1910s because of the high cost of installing pipes and the vehement resistance of the water carriers themselves. However, during the 1920s, the expansion of running water dramatically jeopardized the water distribution territories of the water carriers. I explore the tactics that the water carriers adeptly applied—what I call "the politics of shoulder poles"—to keep the modern Waterworks away from their territories and how, with the support of their well guild, they became water lords. Nevertheless, this process was complicated by factionalism arising from the uneven distribution of running water.

Finally, I examine hygienic water in chapter five, particularly how the state's hygiene mandates affected drinking water and its purveyors during the Nanjing decade (1928–1937). Following the establishment of the Nationalist state in 1928, the state further sought to modernize itself by emphasizing hygienic modernity. As a result, it initiated hygiene programs with the assistance of the Peking Union Medical College (PUMC). John Grant, a public health expert

18 The Beiyang Government, also known as the First Republic of China, was derived from the Beiyang Army.

from PUMC, transplanted the notion of public health to Beijing through Health Demonstration Areas (Health Stations)[19] that allowed the state to better control water, and by extension, the water carriers. At the same time, water carriers used various forms of covert resistance, both collectively and individually, to redefine the city's relationship to water. The Nationalist state was unable to municipalize drinking water and eliminate the water regime, which was finally nationalized by the Chinese Communist Party in the 1950s as discussed in the epilogue.

This book project draws on a large range of local archival research in its examination of the water regime. Gathered through a painstaking and exhaustive exploration of primary sources in Mainland China, Taiwan, the United States, and Japan, three kinds of archival collections have proved especially useful: The principal sources for this book are judicial records of water carriers and supporting police investigations from the Beijing Municipal Archives, which are critical to understand the nature of the water regime. A second set of primary sources is comprised of the drinking water investigations conducted by Peking Union Medical College (PUMC); these records, located in the Rockefeller Archive Center, are crucial to conceptualizing the hygienic water of the state. Lastly, I utilize copies of water contracts from the Qing to Republican periods, which are housed in the Niida Collection of the Oriental Library in the University of Tokyo Library System; and which are central to constructing the territory and order of the water regime.

19 Health Demonstration Areas were areas created by the hygiene department of the PUMC in the east city of Beijing to promote public health through programs for birth, prevention of disease, disinfection of drinking water, and so on.

CHAPTER 1

The Creation of the Water Regime, 1644–1900

1 Introduction

Overshadowed by the splendid palace, fascinating temples, and prosperous shops in the capital of the Qing dynasty (1644–1911), water carriers remained largely unrecognized by travelers and outsiders. However, a Russian mission to Beijing in 1821 bore witness to a group of men supervising wells and water delivery in the streets.

> These wells are kept up by the inhabitants of the province of Chantong [Shandong], who are considered as the best workmen in Peking; and are in fact distinguished from the rest of the Chinese by their lofty stature, bodily strength and open countenance. They bring water into the houses of the rich and the tea houses in Peking, for a certain price, on barrows with two wheels, on each of which are ten pails.[1]

These Shandong water carriers, pushing wheelbarrows loaded with hundreds of pounds of water to distribute, were only the latest water carriers in Beijing's history.

The city had been known for its scarcity of drinkable water since the late imperial period. Wells were considered common-property resources and were maintained by the local communities for drinking and household purposes. As early as the Ming dynasty (1368–1644), communities hired water carriers to wind in and out of narrow *hutongs* (alleyways) to ensure a steady water supply. These water carriers, mainly from Shanxi Province, knew the individual households so well that when the throne selected court ladies from Beijing, the emperors enlisted the water carriers to locate desirable candidates and authorized this extra service through an ordinance.[2] When Li Zicheng (1606–1645), the rebel leader, overthrew the Ming dynasty and occupied Beijing in April of 1644, his men who lusted for women requested that the water carriers hunt for pretty women in the city. The water carriers successfully completed their assignments and won the favor of the rebels; but the influence of the unruly

1 George Timkowski, *Travels of the Russian Mission, through Mongolia to Peking in the Years 1820–1821*, vol. 2 (London, 1827), 38.

2 Shi Xuan, Jiujing yishi, p. 4 (史玄:《旧京遗事》北京: 北京古籍出版社, 1986, 第 4 页。)

 | DOI:10.1163/9789004720800_003

Shanxi water carriers, as accomplices of the insurgents, was immediately diminished after the Manchu defeated the rebels and took over the city in June of 1644.[3]

After the city's capture by the Manchu in the Qing period, residential segregation in post-conquest Beijing was designed to uphold new political and social hierarchies. The Imperial City was located in the center of Beijing. Outside the Imperial City was the Tartar City, or the Inner City, which was reserved as a garrison for Manchu troops charged with the defense of the capital. The Chinese City, or the Outer City, where the Chinese were permitted to live and trade, was to the south of the Inner City.[4] Under the circumstances, the vigilant Manchu regime remade the water delivery system and favored Shandong men, who were loyal followers of the state, as water carriers.

As Lillian Li and Alison Dray-Novey's study on the food security of Qing-dynasty Beijing points out, an essential feature of Beijing's long history as China's imperial capital was its ability to feed its population despite a geographical location distinctly unfavorable to agriculture.[5] Unlike the Ming state, which did not involve itself with the water supply, the Manchu throne, surrounded and outnumbered by the Chinese,[6] considered securing Beijing's drinking water a vital issue in its environmental administration. The reliable availability of clean and safe water was considered to be a necessity for consolidating control and establishing the Manchu regime in Beijing. Accordingly, the Qing state designed water governance that included direct control of supply and indirect control over markets.

In the Qing dynasty, the priority of the state was to secure safe and drinkable water both as subsistence and welfare for the emperor in the Imperial City and the banner units in the Inner City. Consequently, the state designed three water supply systems: Jade Spring water for the emperors in the Imperial City; water

3 Zhang Cixi, *Yanjing fanggu lu*, p. 31–34. (张次溪：《燕京访古录》北平：中华印书局，1934 年，第 31–34 页。)

4 The issue of segregation has been extensively studied. The existing research includes Mark C. Elliott, *The Manchu Way: The Eight Banners and Ethnic Identity in Late Imperial China* (Stanford: Stanford University Press, 2001); Lai Huimin, *Danwen qimin: Qingdai de falü yu shehui* (Taipei: Wunan tushu chuban gongsi, 2007) (賴慧敏，但問旗民：清代的法律與社會 台北：五南圖書出版公司，2007 年。); Liu Xiaomeng, *Qingdai Beijing qiren shehui* (Beijing: Zhongguo shehui kexue chubanshe, 2008) (劉小萌，清代北京旗人社會，北京：中國社會科學出版社，2008 年。)

5 Li Lillian and Alison Dray-Novey, "Guarding Beijing's Food Security in the Qing Dynasty: State, Market and Police," *The Journal of Asian Studies* 58, no. 4 (Nov. 1999): 992–1032.

6 By 1909, Beijing's population had increased to 1.1 million; see Han Guanghui, *Beijing Lishi Renkou Dili* [A Historical Demographic Geography of Beijing] (Beijing: Beijing daxue chubanshe, 1996), 114.

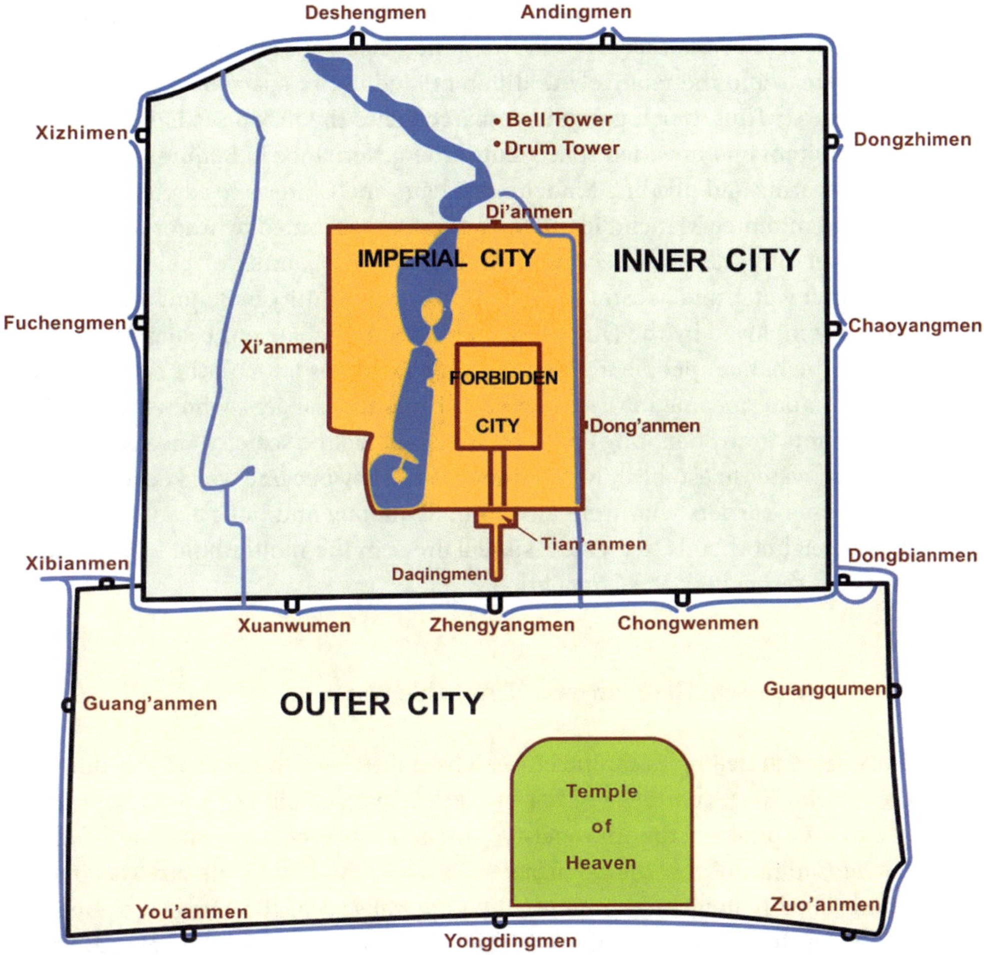

FIGURE 1 A map of Beijing in the Qing period (1644–1911)

from state wells distributed by Shandong water carriers for the banner families in the Inner City; and, in the Outer City, water procured by private water carriers who were hired by residents, without state involvement, to supply water. Thus, the drinking water in the Inner City was a public provision, while water provision for the Outer City was arranged privately by individual residents. The two different water carrier systems—the state carriers in the Inner City and the private carriers in the Outer City—were not only a response to the geography of drinking water but were also constructed by the state to represent the spatial and social dichotomy between the banner and the non-banner people within the capital.

As William Skinner suggests, "modern scholarship tends to emphasize diverse activities that occurred in the same place without distinction among urban areas and the relatively undifferentiated use of space in the preindustrial cities."[7] Thus, much previous research views the urban setting of Beijing as a coherent and unvaried space. But water governance in Beijing was aimed at segregating and dividing Manchu residents and Chinese residents to maintain and reinforce Manchu identity. The water distributed by water carriers in the Inner City was called *erxingzi* water (二性子水), a mix of half fresh and half bitter water, and became a symbol of banner identity in Beijing. The water carriers employed by the Qing state were used to enhance the social integration of the banner people in the Inner City, as well as to mark the segregation of non-banner people in the Outer City. The water carriers, who were mainly immigrants from Shandong Province, manipulated the scarcity and security of drinking water to establish water enterprises. They became well keepers, then hired water carriers who were also from Shandong and built a water regime in the neighborhoods to create a spatial order in the multiethnic Beijing that persisted throughout the Qing period.

2 Bitter Sea: The Scarcity of Drinkable Water

The choice of Beijing as capital city can be traced back to the Yan State during the Warring States period (475 BCE–221 BCE). In 1153, Beijing became Zhongdu (Central Capital) of the Jin dynasty (Jurchen Jin, 1115–1234), and then Dadu (Great Capital) of the Yuan dynasty (1271–1368). After this, the city was once again called Beijing and remained the imperial seat of the Ming (1403–1644) and Qing dynasties (1644–1912). Despite its prominence as the capital, Beijing consistently suffered from water shortages. Located on the semi-arid North China Plain, it is one of the few major capitals of the world that is not located on a river of considerable size. The Yongding River to the southwest was intended to serve as Beijing's source of water. However, it turned out to be a source of flooding disasters because of constant silting, which forced the city to locate itself at a reasonable distance from its main river. Moreover, Beijing had a continental climate—several monsoons and long, dry, cold winters—that further aggravated the shortage of water. This situation barely changed despite repeated hydraulic projects undertaken throughout the city's history.[8]

---

7 G. William Skinner, "Introduction: Urban Social Structure in Ch'ing China," in G. William Skinner, ed., *The City in Late Imperial China* (Stanford, CA: Stanford University Press, 1977), 3–5.

8 Hou Renzhi, "Beijing tushi guocheng zhong de shuiyuan wenti," in Hou Renzhi, *Lishi dilixue de lilun yu shijian* (Shanghai: Shanghai renmin chubanshe, 1979), 296–303.

The open water system that included rivers and lakes failed to supply plentiful drinking water for the city; it was also highly susceptible to the impact of climatic factors such as floods in the summer and ice in the winter. As a result, Beijing relied on wells instead of rivers and lakes to procure drinking water for its population. Fortunately, an extensive underground water table lay beneath the city, from which water could be obtained by digging wells. Thus, over time, Beijing developed a tradition of using wells for its drinking water. The archeological evidence indicates that the history of digging wells in Beijing can be traced back to the Warring States period.[9]

The groundwater from these wells, which were usually thirty feet in depth and consisted of nothing but a brick- or rubble-lined shaft sunk into the uppermost layer of soil, was called "bitter water" because of its disagreeable taste. Pure and palatable water, called "sweet water," was rare. From a geological point of view, the plain upon which Beijing sits was built of an extremely mixed set of layers of loess, clay, sand, and gravel because of the constant shifting of streambeds. The fluviatile formation and heterogeneous bedded character of the plain meant that the productive and good water zone lay at a depth of around 200 feet.[10] But the hand-dug drilling technology of the preindustrial period limited access to this deeper and better-tasting groundwater. Accordingly, the water sourced from the shallow wells in Beijing contained a good deal of calcium, magnesium, and other dissolved minerals, which dramatically contributed to its hardness and brackishness. Even worse, the city's dense population caused the accumulation of nitrates and pollution, which further deteriorated the underground water. As a result, historically, Beijing was notorious for its salty and turbid water.

Grievances about salty drinking water are consistently reflected in the writings of city dwellers. For instance, the walled city of Dadu had a population of 418,000 in 1270, which rose to 952,000 plus another 2.08 million residents in the surrounding region by 1327.[11] Wang Yun, a notable writer and official who resided in Dadu, complained of the water's quality in the following terms: "Due to the infiltration of decomposed garbage, the drinking water is salty and less than 2 percent of the wells in Beijing have pure water."[12] In the Ming dynasty,

9 Zhang Xiande, "Shuijing de lishi ji Beijing chutu shuijing gaishu," in Su Tianjun, ed., *The General History of Wells in Beijing* (Beijing: Beijing chubanshe, 2000), 274–275. (張先得《水井的歷史及北京出土水井概述》，蘇天鈞編：《北京考古集成》，北京：北京出版社，2000年，第1卷，第274–275頁。)

10 Bailey Willis, *Research in China*, vol. 1, pt. 1 (Washington: Carnegie Institution of Washington, 1907), 196.

11 Han Guanghui, *Beijing lishi renkou dili* [A Historical Demographic Geography of Beijing] (Beijing: Beijing daxue chubanshe, 1996), 137.

12 Wang Yun, *Qiujian ji*, vol. 36, Xinjing ji (王惲，秋澗集，卷36，新井記。)

city inhabitants continued to recount this problem. Liu Ruoyu, a eunuch who was a lifelong resident of Beijing, grumbled that the majority of well water was saline and hard. When used to infuse tea, if teacups and teapots were not cleaned for three days, they would be encrusted with a considerable deposit of lime.[13] Nevertheless, city dwellers had no choice but to rely on this water as a basic resource.

The dominance of bitter water was even imbedded in the city's collective memory. One well-known legend is about Gao Liang's race for water (高亮趕水). In this legend, the Dragon King, the god who oversaw water and weather in Chinese folk religion, intended to dry out Beijing in revenge for humans' occupation of his territory. He packed all of Beijing's water into two pouches, fastened them to a cart, and left. A warrior named Gao Liang, volunteering to recover the water from the Dragon King, raced toward the cart and thrust his spear into the water pouches. But in his rush he made a grave mistake. He only speared the water pouch on the right side. But the left pouch contained sweet water while the right pouch contained brackish water. Later on, Beijing had many salty wells, which was attributed to Gao Liang slitting the wrong water pouch. As for the left water pouch still on the Dragon King's cart, it was used to fill Jade Spring Hill and ever since the water gushing from it has been very sweet and tasty.[14] The Gao Liang story reflects the dire historical reality that sweet water in Beijing was always rare.

Thus, long before the Ming and Qing dynasties, the predominance of bitter water, with its salty and alkaline taste, had won Beijing another nickname: "Bitter Sea" (Ku Hai). In 1644, the Manchu achieved the complete conquest of China after decades of campaigns and established Beijing as its capital. To his dismay, the prince regent Dorgon (1612–1650) of the new Qing dynasty discovered that the drinking water was so unpalatable that he even proposed relocating the capital to Shenmuchang (神木廠), east of Beijing, in order to be closer to good drinking water. His plan was eventually aborted because of the tremendous cost.[15] Nevertheless, securing good-tasting drinking water in the bitter sea was the first issue the Manchu regime had to address, while procuring "safe water" became the other pressing task on the new state's agenda.

---

13 Liu Ruoyu, *Zhuo zhongzhi*, vol. 17 (劉若愚，酌中誌，卷十七。)

14 Chan, Hok-Lam, *Legends of the Building of Old Peking* (Hong Kong: The Chinese University of Hong Kong, 2008), 226–231.

15 *Qingshilu*, vol. 44, 2. (清世祖實錄卷44，順治六年五月癸亥條，北京：中華書局，2008年，第3冊，第2頁。)

3 Safe Water: The Security of Drinking Water

Facing new challenges to its precedents, the Qing state sought to grapple with the security of the city's drinking water through the reorganization of the environment, including water resource collection and delivery. The dangers involving water security that the Manchu were concerned about were wars and social disturbances; accordingly, the Manchu made every effort to minimize the risk of water poison and contamination imposed by their foes. Originating from sedentary agriculture, the Manchu had developed the tradition of using wells in their settlements as their chief water sources. So crucial were these wells to their daily life that the Manchu worshiped a well deity (井神), a practice that continued in their conquest of China. On each winter solstice, the Manchu court would make sacrificial offerings to the well deity, along with a bilingual prayer to the deity to ask that the water be clear and clean.[16]

The Manchu had also learned lessons for well protection from their war with the Han Chinese in the early seventeenth century. Manchu archives repeatedly recorded and highlighted cases of wells poisoned by the Chinese during the Ming dynasty.[17] For instance, in May 1621, when Nurhaci (1559–1626), the ruler of the Manchu, was making an inspection in Haizhou, local military officials succeeded in hunting down eight Chinese men who were plotting to poison the wells. So furious was Nurhaci that he forced the Chinese to swallow the poison they carried and ordered local officials to attentively guard the wells.[18] Following their ancestors, Manchu emperors like Kangxi (康熙, 1654–1722) were extremely vigilant regarding the security of drinking water. Leaving wells unguarded, when the lives of all depended upon the purity of the water, was an offense comparable to treason against the state. During an expedition to Mongolia in 1696, the Kangxi emperor found that two officers assigned to this duty

16 Evelyn Rawski, *The Last Emperors: A Social History of Qing Imperial Institutions* (Berkeley: University of California Press, 1998), 268.

17 The fear of well poisoning is a global phenomenon and accusing one's enemies of poisoning water supplies portrays them symbolically as depriving people of what is most essential for sustaining life. It thus speaks directly to the collective fear of death in the minds of ordinary people and provides a ready-made vehicle for mobilizing people to defend themselves and the state. See Paul A. Cohen, *History in Three Keys: The Boxers as Event, Experience and Myth* (New York: Columbia University Press, 1997), 167–169.

18 Zhongguo diyi lishi danganguan, *Manwen laodang* (Beijing: Zhonghua shuju chubanshe, 1990), 206. (中國第一歷史檔案館，中國社會科學院歷史所譯註，滿文老檔 北京：中華書局 1990 年，太祖皇帝天命六年五月。)

had not remained at their posts. He ordered them beaten, divided their horses and possessions among the soldiers, and sent them into exile in the north.[19]

Once settled in Beijing in 1644, the Manchu minority understood that they were surrounded by potentially hostile Chinese. Securing clean and safe water for the emperor became a priority of the Manchu state. The best source was the celebrated Jade Spring water, located ten miles northwest of Beijing. The Jade Spring was noted for its sweet water and had been an attraction in Beijing since the Jin dynasty. Shen Bang, a local magistrate during the Ming dynasty, depicted Jade Spring in his jurisdiction as follows:

> Jade Spring, located thirty *li* away from Beijing, gushes out of the rock of the Jade Spring Hill. The spring jumps down from the hill in the form of a waterfall into a pond. The water is extremely pure, cold, and odorless. It is one of eight attractions in Beijing, namely, The Rainbow of Jade Spring [玉泉垂虹].[20]

This tasteless and colorless water was the perfect drinking water source for the palace. During the Kangxi period, the ambitious emperor built the Jingming Garden (靜明園) to enclose Jade Spring. In this way the spring water was monopolized by the throne and consumed exclusively by the palace; one hundred and three bondservants and eunuchs were assigned to guard and deliver this rare water.[21] During the reign of Qianlong (乾隆, 1711–1799), the emperor reevaluated all the renowned springs across his empire and ranked Jade Spring water as the First Spring under Heaven (天下第一泉).

To procure drinking water for the emperors, the regime institutionalized the Jade Spring water delivery system. The conveying of this water to the palace was handled by the Imperial Household Department.[22] The department sent three officials to regularly oversee and deliver spring water to the palace.[23]

19 Eloise Talcott Hibbert, *Jesuit Adventure in China: During the Reign of K'ang Hsi* (New York: E. P. Dutton and Company, 1941), 220.

20 Shen Bang, *Wanshu zaji*, in Zhongguo kexueyuan tushuguan bian, *Xijian zhongguo defangzhi huikan* (Beijing: Zhongguo shudian, 2007), 29. (沈榜，宛署雜記 卷四，中國科學院圖書館選編，稀見中國地方誌匯刊 北京：中國書店， 2007 年。)

21 *Qianlongchao qingding daqing huidian zeli*, vol. 160, 46. (乾隆朝钦定大清会典则例，卷一百六十，中国第一历史档案馆，大清历朝实录，第 46 页。)

22 Wu Zhenyu, *Yangjizhai conglu* (Beijing: Zhonghua shuju, 2005), 302. (吳振棫，養吉齋叢錄北京：中華書局 2005 年, 卷 24。)

23 Gugong bowuyuan, ed., *Qingding neiwufu zeli erzhong* (Haikou: Hainan chubanshe, 2000), 293. (故宮博物院編，欽定內務府則例二種 海口：海南出版社，2000 年, 第2冊, 掌關防管理內管領事務處 卷1, 取玉泉水，第 293 页。)

FIGURE 2 Water cart for the imperial family
SOURCE: ILLUSTRATION DRAWN BETWEEN 1764 AND 1772, IN JOSEPH BRETON, *CHINA: ITS COSTUME, ARTS, MANUFACTURES &C*, VOL. 1 LONDON: J. J. STOCKDALE, 1813, P. 80.

Early every morning, an imperial water cart loaded with large, cube-shaped tin vessels, decorated with a yellow imperial flag, pulled by a donkey, and escorted by twelve soldiers, would go out of the city through Xizhi Gate (西直門) to deliver this water to the emperor once per day. The department also stored sixty tin tanks for carrying spring water.[24] As one Jesuit missionary in Beijing observed: "Very great attention is paid to taking the water from that part of the stream which is the clearest. The carts are followed by eunuchs and confidential officers, who never lose sight of them. The least neglect in what relates to the daily consumption of the emperor and princes, would be liable to very rigorous punishment."[25] This delivery system persisted well into 1924, the year that the last emperor, Puyi (溥儀, 1906–1967), was expelled from Beijing.[26]

24 Chinese First Historical Archives, *Wei chengming qingling shuichunchu wang jingmingyuan qushui suoxu xiguan shi*, No. 05-08-009-000162-0010. (中國第一歷史檔案館：为呈明请领水轮处往静明园取水所需锡罐事，05-08-009-000162-0010，道光三年七月二十一日。)

25 Joseph Breton, *China: Its Costume, Arts, Manufactures &c*, vol. 1 (London: J. J. Stockdale, 1813), 80.

26 Fu Jia, "Qinggong huiyi," in *Wanqing gongting shenghuo jianwen* (Beijing: Wenshi ziliao chubanshe, 1982), 12. (溥佳，清宮回憶，中國人民政治協商會議全國委員會文史資料研究委員會編，晚清宮廷生活見聞 北京：文史資料出版社，1982 年。)

The daily amount of Jade Spring water allotted for the emperor and empress was twelve jars each. The emperor relied exclusively on this water, even when he was on an inspection tour. Royal inspection tours required both logistical and political management, and this in itself became a measure of the court's capacity for disciplined and proper governance.[27] When the Qianlong emperor was on his trips to Jiangnan, the Imperial Household Department would deliver Jade Spring water every day through the courier station system at enormous cost.[28] Apart from drinking purposes, Jade Spring water was also used to brew the imperial Jade Spring wine (玉泉酒).[29]

Jade Spring water was so closely associated with the emperors that it became a symbol of royal power. Accordingly, the emperor would give this water to his favored ministers. For example, in 1713 when the Grand Secretary Li Guangdi (1642–1718) was sick, the Kangxi emperor advised him to not drink water from the city's wells. Instead, for nearly a month, the emperor reserved a portion of his share and bestowed two jars of spring water each day to cure Li's diarrhea.[30] Another emperor, Guangxu (光緒, 1871–1908), rewarded Weng Tonghe (1830–1904), the emperor's tutor, a jar of Jade Spring water.

While Jade Spring water was the principal source of the emperors' drinking water, its delivery system was susceptible to weather, especially to storms and heavy snow that would obstruct the road to Jade Spring Hill. When delivery was halted for a few days, the emperors would use water from wells in the palace. Inside the palace, the Dapao Well (the Great Kitchen Well) in front of the imperial kitchen was a famous sweet-water well that supplied drinking water for the palace. Eight eunuchs took charge of carrying water from it each day. The Dapao Well was so significant that it was the site where the palace performed its worship of the well deity on each winter solstice.[31]

---

27 Michael G. Chang, *A Court on Horseback: Imperial Touring & the Construction of Qing Rule, 1680–1785* (Cambridge, MA: Harvard University Press, 2007), 114–159.

28 Chinese First Historical Archives, *Zouwei jiaxing Shandong dai yuquanshui*, No. 05-0283-018. (中國第一歷史檔案館檔案奏爲駕幸山東帶玉泉水用珍珠泉林水事，05-0283-018，乾隆三十五年十二月初二日。)

29 Wu Zhengge, *Manqing shisu yu qinggong yushan* (Shenyang: Liaoning keji chubanshe, 1988), 376. (吳正格編著，滿清食俗與清宮禦膳 沈陽：遼寧科學技術出版社 1988年，膳食檔。)

30 Chinese First Historical Archives, ed., *Qingdai dangan shiliao congbian, 9* (Beijing: Zhonghua shuju, 1983), 23–24. (中國第一歷史檔案館編，清代檔案史料叢編 第9輯，北京：中華書局 1983 年，第 23–24 頁。)

31 Ji Yun: *Yuewei caotang biji.* (紀昀：閱微草堂筆記，第十三卷 槐西雜誌三)

As a consequence, a miniature temple of the Dragon King was erected in front of this well.[32]

According to Huang Jian, an imperial tutor (侍講学士) and a master of water appraisal in the Ming dynasty, while the best water outside Beijing was from the Jade Spring, the best water within Beijing was from the Dapao Well.[33] Thus, the Manchu palace monopolized the two best water sources in Beijing. Of course, the Dapao Well was not the only well in the Imperial City. It had a total of seventy-two wells, as indicated in the Qianlong Map of Beijing. All this well water was limited to use within the palace. In the Jiaqing period (嘉慶, 1796–1820) some personnel surreptitiously carried water away from a horse-stable well near Donghua Gate. The chief officer of gate custodians was deprived of his salary for one year as a punishment for his negligence.[34] In general, using the service of bondservants and eunuchs, the imperial city instituted a separate water delivery system to access safe and pure water, including Jade Spring and well water, which was independent from the water carrier systems of the Inner City and the Outer City that are the topic of this book.

4 State Providers: Water Supply in the Inner City

The Manchu state's objective in its water administration was not only to supply its rulers; it also sought to provide water to its followers in order to preserve the security of the capital. Thus, the real challenge of the Manchu state was to procure safe and drinkable water for its numerous banner troops and their families living in Beijing. The Manchu created the "eight banners" at the turn of the seventeenth century, and these eventually became the organizational vehicle for conquering China.[35] The banners formed the social and military units of the Manchu. When they arrived in Beijing in 1644, the Han Chinese

---

32 Chinese First Historical Archives, *Wei jieyong yingzaosi yinxin qingling chaixiu yushanfang yuannei shishuijing yuqiang*, No. 05-08-006-000610-0063. (中國第一歷史檔案館檔案，为借用营造司印信请领拆修御膳房院内食水井宇墙等项所需工料银两事，05-08-006-000610-0063，道光十六年九月初七日。)

33 *Guochao xianzhenglu*, vol. 20. (國朝獻徵錄·卷20 黄谏。)

34 Chinese First Historical Archives, *Fenggongye zou wei Donghuamen nei jingshui xiansan rending zi tiaoqu, jiang shizhi zhinian neiguanling Chang Kai deng fafengshi*, No. 05-0496-001, July 1, 1802. (中國第一歷史檔案館 內務府奏案 05-0496-001 奏折，豐公爺奏爲東華門內井水閑散人等自挑取，將失職值年內管領常愷等罰俸事，嘉慶七年七月初一日。)

35 The eight banners were Plain Yellow and Bordered Yellow, Plain White and Bordered White, Plain Red and Bordered Red, and Plain Blue and Bordered Blue. The Yellow Banners were stationed north of the city, the White Banners were stationed in the northeast

inhabitants in the north of the city were driven from their homes to provide dwellings for the banner units. The evicted Chinese were relocated to the adjacent city in the south, named the Outer City, which was divided by high walls from the Inner City.

One factor in particular should be mentioned for understanding this relocation: geographically, the quality of drinking water in the Inner City was better than in the Outer City. As one local maxim put it: "The best tea of Beijing is in the commercial district of the Outer City while the best water is in the Inner City" (南城茶葉北城水). Moreover, the internal hierarchy of the eight banners was reflected in each banner's distance to water. The three upper banners under the direct command of the emperor were assigned to the north part of the Inner City where they had greater access to good water as compared to the five lower banners in the south.[36]

The Inner City was reserved as a garrison for the banner troops charged with the defense of the capital. Although engaged with the Outer City for services such as trade and entertainment, the Inner City functioned as a self-sustaining city. It is estimated that the population of banner troops and their families within the Inner City in the early Qing numbered about 400,000.[37] The Qing state's supervision of banner life was deeply bureaucratized and paternalistic. Manchu residential areas were divided into eight quarters, each under the jurisdiction of one banner. Each quarter was further subdivided into about sixty companies (佐領) whose importance was enhanced at the expense of kinship units. Employment was allocated, stipends distributed, and residence determined primarily through approximately 600 banner companies.[38]

Beginning in 1644, the Qing rulers provided a variety of extensive special public services exclusively for the privileged people in the Inner City.[39] One of the important services offered by the state was drinking water. It was not

---

and northwest of the city, the Red Banners occupied the district in the west, and the Blue Banners were quartered in the extreme south of the Inner City.

36 A hierarchy existed within the eight banners. The three upper banners (Plain Yellow, Bordered Yellow, Bordered White) were under the direct control of the emperor, while the five lower banners were responsible to imperial princes. See Mark C. Elliott, *The Manchu Way: The Eight Banners and Ethnic Identity in Late Imperial China* (Stanford, CA: Stanford University Press, 2001).

37 Han Guanghui, *Beijing lishi renkou dili* (Beijing: Beijing daxue chubanshe, 1996), 114–29. (韓光輝, 北京歷史人口地理, 北京：北京大學出版社，1996年，第114–129頁。)

38 Mark Elliott, *The Manchu Way: The Eight Banners and Ethnic Identity in Late Imperial China* (Stanford, CA: Stanford University Press, 2001), chap. 4.

39 Susan Naquin, *Peking: Temples and City Life, 1400–1900* (Berkeley: University of California Press, 2000), 371.

uncommon for the state to remind its followers that they received their water through imperial beneficence.[40] The state's approach was to reconstruct old wells and dig new wells for the banners. To further heighten the security of drinking water, all the wells in the streets of the Inner City were state-owned and the state placed limitations on private wells. Ordinary people were forbidden to drill wells without permission from the state, although a powerful group of banner elites in the Inner City received special privileges to drill wells in their own compounds and mansions.[41] For example, the wife of Bo Qishan from the Aisin Gioro ruling clan owned a compound with 115 rooms and two sweet-water wells.[42] Politically, the state sought to monopolize the sources of safe water, while aiming to minimize external threats and thereby facilitate regulation. Economically, it was unfeasible to drill wells for individual banner families because the construction of hand-dug wells for accessing groundwater was a time-consuming and costly enterprise.[43]

The ban on private wells was implemented so effectively that state-owned wells became the chief sources of drinking water in Beijing's Inner City and the number of wells remained remarkably consistent. For example, in the Qianlong period the number of wells, as identified on *The Map of Beijing*, was 676; it increased only slightly to 694 during the one-hundred-year span of the Guangxu period.[44] The ban against private wells was not lifted until the last decade of the Qing dynasty, because the emperor continued to maintain tight control over well drilling in the Inner City. For instance, in 1902, the Guangxu emperor ordered the Nine Gates Infantry Commander who oversaw Beijing's defenses and security to dig more wells for the dense population inside the Inner City.[45]

---

40 Matthew Keith Malott, *Nomine Caesaris: An Examination of the Propagandistic Functions of the Aqueducts of Rome during the Early Empire* (Classical Association of Canada Essays, 2004).

41 The study of real estate transactions in Qing Beijing suggests that well ownership by upper-class banner households was common. See Zhang Xiaolin, *Qingdai Beijing chengqu fangqi yanjiu* (Beijing: Zhongguo shehui kexue Chubanshe, 2000), 233–250. (張小林，清代北京城區房契研究，北京：中國社會科學出版社，2000 年，第 233–250 页。)

42 Liu Xiaomeng, Qingdai Beijing qiren shehui (Beijing: Zhongguo shehui kexue chubanshe, 2008), 144. (刘小萌：《清代北京旗人社会》北京：中国社会科学出版社，2008 年，第 144 页。)

43 Li Wan, *Jingcheng guji kao.* (勵萬，京城古跡考 記滿井雲 "北方地脈高厚，或掘井數仞，猶不及泉。)

44 Zhu Yixin, *Jingshi fangxiang zhigao* (Beijing: Beijing guji chubanshe, 1982). (朱一新，京师坊巷志稿，北京：北京古籍出版社，1982 年。)

45 *Qingshilu,* vol. 58, 617. (清實錄 光緒 28 年 1902 年 壬寅 六月十五日 ，第 58 冊，第 617 頁。)

Apart from serving banner families, street wells also slaked the thirst of city animals such as horses, camels, and donkeys.

To further protect and facilitate the water supply, the state assigned those who had previously served as cooks in the logistics department of each banner company to watch over the wells within the territory of the companies. The cooks who guarded these state-owned wells as well keepers originated from Shandong Province. In Late Imperial China, Shandong was a densely populated and disaster-prone province in north China, which created instability and forced population movements. The vast and sparsely populated area of Manchuria was the conventional destination for Shandong migrants. During the Ming period (1368–1644), the Shandong peninsula and Liaodong peninsula were under the same jurisdiction. The sea route between these two regions was less than 120 miles, which allowed Shandong people on the Shandong peninsula, especially peasants from Dengzhou Prefecture (登州府), to easily flood into Manchuria. In the early seventeenth century, thousands of Shandong migrants were registered in the newly created banner system by Nurhaci and served as infantry support for the banner people. As a reward, the Manchu state entrusted the state wells in the Inner City to the hands of Shandong men and granted them the right to lease wells for indefinite periods and to pass these leaseholds on to their sons.[46]

As well keepers, these trustworthy Shandong men built shelters near the wells to ensure their security and purity, for which they received state stipends. The well keepers who took on the obligation to protect water from misuse and pollution were also engaged in water distribution to the banner people's doorsteps and developed a water carrier system in the Inner City that met several needs. First, the number of wells was limited due to the cost of well drilling and most banner households lived a certain distance from wells with drinkable water. Second, the majority of bannermen had taken up permanent positions in the banner armies and the responsibility of procuring daily water fell to women and children, who were unable to endure the stress of carrying heavy water over considerable distances. More importantly, the water quality of street wells was improved by frequent drawing while the standing water in private wells in the compounds was comparatively brackish. As a result, people became accustomed to using water from the street wells.[47] Water distributed by well keepers became necessary and popular in banner families.

46 Zhang Cixi, *Yanjing fanggu lu* (Beiping: Zhonghua yinshuju, 1934), 31–34. (張次溪，燕京訪古錄 北平：中華印書局 ，1934 年，第 31–34 頁，擔水夫考。)

47 John Dudgeon, *Diet, Dress and Dwellings of Chinese in Relation to Health* (London: William Clowes and Sons, Limited, 1884), 292–293.

Water carriers delivered both "bitter water" and "sweet water"; the former was used for washing, sprinkling, and gardening purposes while the latter was used for drinking and cooking. Although water from the state wells was free, the service of drawing and delivering water had a price. Well keepers withdrew water from wells and then distributed it according to the boundaries of company residential quarters. Thus, the water distribution territories originated from the living patterns of the banner people. Each individual well keeper possessed the exclusive right to distribute water within his territory, and each territory had clear boundaries. One keeper typically served thirty to sixty families.[48] Tan Qian, a historian who resided in Beijing for two and half years in the 1650s, observed that all the streets and lanes in Beijing had men who distributed sweet water for particular areas, and these areas couldn't be encroached upon by others.[49]

In everyday life, well keepers who were assigned to good-quality water wells saw more opportunities to provide convenient service for profit and stretched their tentacles into the territories of bitter-water well keepers. A lawsuit during Kangxi's reign illustrates the expansion of water distribution territories. One sweet-water well keeper subleased his wells and distribution territory of twenty households to another well keeper. When the lease expired five years later, the two had a dispute because the number of households within the territories had increased to forty in the period. The lessee refused to return the extra twenty households to the lessor.[50] This example showcases the expansion of water distribution territories and the appeal of sweet water for the banner families in the Kangxi period, and it also hints at the ongoing conflicts and negotiations regarding the boundaries of water distribution territories in the Inner City of Beijing.

When the water business expanded, the Shandong well keepers created a "well house" system, named after the protective structures built around wells. These well house systems were usually composed of a number of wells, a well keeper with one or multiple water distribution territories, and a few water carriers. In the daily operation of the enterprise, the well keeper as the owner of a well house provided accommodations and salaries to his water carriers, who drew water from wells with hand buckets, stored the water in cisterns in

48 For example, one water distribution territory in Xidan in 1882 had thirty-seven families. See water contract no. 10–294, housed in the Niida Collection, Tokyo University, Oriental Library.

49 Tan Qian, *Beiyou lu* (Beijing, Zhonghua Shuju), 1960, 312. (談遷，北遊錄，紀聞上，北京：中華書局，1960 年。)

50 Lu Yi, *Xuncheng suoji*, 17–18. Department of Rare Books, Nanjing Library, China. (陆毅，巡城瑣紀，南京圖書館古籍部特藏。)

FIGURE 3 Water carriers and a well on a street in Beijing, shown in *Drawings of the Sixtieth Birthday Process of Kangxi Emperor* (萬壽盛典圖) in 1713, by Wang Yuanqi.

front of the wells, and then delivered the water either in wheelbarrows or on shoulder poles within the distribution territories of the well house. In addition, the well houses built tiny temples to the Dragon King and worshipped the well deity on the first and fifteenth day of every Chinese lunar month, which was believed to protect the cleanliness and plentifulness of well water (see fig. 3).

Through these water enterprises, the presence of the water carriers became more visible in the everyday life of the Inner City. Even the *Drawings of the Sixtieth Birthday Procession of the Kangxi Emperor* (萬壽圖盛典圖) portrayed water carriers, a well with a trough, and a temple in the street.[51]

Within the well house system, the Shandong well keepers became owners and managers, and generally preferred to hire carriers from their native places. Thus, the majority of water carriers came from Dengzhou Prefecture where the well keepers originated. For example, water carriers involved in seven criminal cases extracted from Penal Memorials (刑科題本) in the Qing period all

51 Drawings of the Sixtieth Birthday Procession of the Kangxi Emperor, *Wanshou shengdian tu* (Beijing: Xueyuan chubanshe, 2001). (萬壽盛典圖 1717 年, 北京：學苑出版社，2001 年。)

came from Dengzhou.[52] The prevalence of Shandong water carriers was vividly depicted in one bamboo twig poem: "Shandong water carriers will have eternally thriving business, unless all wells in Beijing are exhausted."[53] Thus, with the patronage of the Manchu state, Shandong well keepers created the well house system to ensure a supply of safe drinking water in the Inner City.

5 Private Providers: Water Supply in the Outer City

If water in the Inner City was conceptualized as a common good, water in the Outer City was a commodity; the two different water supply models reinforced the structures of social inequality, and water become interconnected with wider issues of social cohesion and political legitimacy. Settling in Beijing in 1644, the Manchu state vacated the Inner City by relocating its former Chinese inhabitants to the Outer City. Although the two cities were more or less equal in size and shape, the equality ended there: The Manchu enjoyed all the privileges that fell to the conqueror, while their neighbors were discriminated against in matters of taxation and public service. The accessibility and availability of drinking water was employed as an apparatus to define racial boundaries and social hierarchy. The Manchu state established a water supply system for the banner families, while the supply of drinking water for the non-banner householders was not a state concern.

The dynamics within the Outer City demonstrate the issues of drinking water quite differently. This intentional neglect of the Outer City had consequences for the city's development, and the pattern of water distribution arose in response to the absence of the state. While the Qing state focused on the Inner City to heighten the security of the banner people, the drinking water for the Outer City was left in the hands of the Chinese. Initially, water was still treated as a common good and no person could be denied water for their domestic needs. Since a water supply infrastructure was not provided within the Outer City, obtaining water was regarded as an individual or domestic responsibility. As a Korean emissary to Beijing in the early Qing observed, there

52 Guo Songyi, "Qingdai Beijing de Shandong yimin," *Zhongguo shi yanjiu*, no. 2 (2010): 27–46. (郭松义：《清代北京的山东移民》，《中国史研究》，2010 年第 2 期，第 27–46 页。)

53 De Shuoting, *Caozhu yichuan*, in Lu Gong, ed., *Qingdai Beijing zhuzhici* (Beijing: Beijing chubanshe, 1962), 43. (得碩亭：《草珠一串》，路工：《清代北京竹枝詞》（十三種），北京：北京出版社，1962 年。)

were two wells on the roadside and people came to collect water by donkey or shoulder pole.[54]

The fact that the large population of Chinese were forced to relocate from the Inner City to the Outer City made access to drinking water a compelling problem. Even worse, the Outer City's water was inferior to the Inner City's water, as Emperor Kangxi himself acknowledged in his edict.[55] In the Outer City, competition for access to drinking water was more intense and its residents engaged in various activities to procure their daily water. In contrast to their counterparts in the Inner City, residents in the Outer City had to dig their own wells.[56] However, as noted earlier, well construction in Beijing was a costly and time-consuming process. As a consequence, the community collectively dug a number of public wells in the streets. In addition, temples provided water to residents from some of their wells. For example, Fayuan Temple (法源寺) provided water to its devotees from a garden well with clean water.[57]

The issue of drinking water in the Outer City was further complicated by the fact that some of the good wells in the Outer City were appropriated by local strongmen and this led to frictions within the neighborhoods. For example, in a Muslim community in the Outer City, a pampered bondservant of a Manchu princess owned a sweet-water well named Zhaojia Well (赵家井) from which Muslim residents obtained their daily water. Around the well was the well owner's vegetable garden, but residents would trample the garden when extracting water from the well. Irritated at this rude behavior, the well owner would catch the offenders who damaged his property, tie them to a tree, and beat them ruthlessly. This caused much tension between the well owner and the Muslim community.

In the twenty-second year of the Kangxi reign (1683), a man named Ma Er unintentionally trampled the vegetable garden. The bondservants of the well

54 Kim Ch'ang-ŏp 金昌業, *Yŏnhaeng ilgi* 燕行日記 [Diary of a China trip], in Im Ki-jung 林基中, ed., *Yŏnhaengnok chŏnjip* 燕行錄全集 [A collection of travelogues written by Korean envoys who traveled to China between the 13th and 19th centuries], 100 vols. (Seoul: Tongguk Taehakkyo Chulpanbu, 2001), 32:16.

55 Gugong bowuyuan tushuguan bian: *Zhanggu congbian*, vol. 2, 12. (故宮博物院圖書館掌故部，掌故叢編，第二輯 1928 年, 清聖祖論旨二。)

56 For example, Liu Ba, who lived on Ox Street during the Kangxi period, was a well driller. Residents who intended to dig wells would contract him. See *Beijing niujie zhishu-Gangzhi* (Beijing: Beijing chubanshe, 1990), 18. (北京市政協文史資料研究委員會等編，北京牛街誌書-岡誌，北京：北京出版社，1990 年。)

57 Longwang pusa lingjing ji (1772), *Beijing tushuguan cang zhongguo lidai shike taben huibian*, vol. 73 (Zhengzhou: Zhongzhou guji chubanshe, 1990), 73. （《龙王菩萨灵井记》（清乾隆三十七年），北京图书馆金石组编：《北京图书馆藏中国历代石刻拓本汇编》，郑州：中州古籍出版社，1990 年，第 73 册，第 73 页。）

owner rushed out and tied Ma Er to a tree, beating him all day long, while the stubborn Ma Er abused them incessantly until he died that night. Furious at the death of his elder brother, Ma San knelt in front of the mosque to enlist help for revenge. The Muslims who attended the mosque immediately appealed to more than one hundred followers to take revenge on Ma Er's behalf. They overthrew the well owner and broke into his house, assaulted women, dismantled his house, looted his belongings, and dumped his clothes into the well. Terrified at this violence, the well owner turned to the city authorities and had four Muslim men arrested. One of the defendants was unable to stand the torture and confessed that he had more than forty accomplices. However, most of those named were innocent Muslim residents with whom the confessor did not get along, while the real culprits remained untouched. In the end, thirty people were sentenced to exile three thousand *li* from Beijing.[58]

The social inequality of drinking water presented the enterprising Shandong water carriers in the Inner City with opportunities to seek profits in the Outer City by introducing the well house system to the Outer City. In the Ming dynasty, Chinese residents in Beijing hired water carriers who hailed from Shanxi Province to deliver water, but they were not attached to any service territories. After 1644, Shanxi water carriers retreated to the Outer City with their Chinese customers and continued to distribute water there. However, two main channels led to Shandong water carriers penetrating the Outer City. The first channel was the relocation of the banner people. The increasing banner population compelled the Manchu state to relocate extra banner people to the Outer City. In 1695, Emperor Kangxi built two thousand rooms for banner people on the training grounds of the Plain Blue Banner at Chongwen Gate (崇文門) and the Border Blue Banner at Youan Gate (右安門), both in the Outer City, to resettle surplus people from these two banners.[59] As a rule, the state assigned Shandong water carriers to the wells that the state had dug for the banner people. These water carriers would then organize well houses and run their water businesses in the Outer City as they did in the Inner City.

The second channel that brought Shandong water carriers to the Outer City was the networks that existed between them and Chinese bureaucrats. In the Qing period, Chinese officials were clustered in the Outer City, especially south of Xuanwu Gate (宣武門). Shandong water carriers sought the patronage of these Chinese bureaucrats in order to expand their business into the Outer

58 *Beijing niujie zhishu-Gangzhi* (Beijing: Beijing chubanshe, 1990), 28–30. (北京市政協文史資料研究委員會等編，北京牛街誌書-岡誌，北京：北京出版社，1990 年。)

59 *Baqi tongzhi chuji,* vol. 33, 438–439. (八旗通誌初集 卷23 營建誌一，吉林：東北師範大學出版社，1985 年，第 438–439 頁。)

City.[60] For example, in the Qianlong period, Wang Fu, an official who was responsible for arranging state banquets (光祿寺卿) and also a native from Dengzhou Prefecture in Shandong Province, was accused of usurping water territories along with his townsmen and competing with civilians. As a result, the emperor dismissed him from his post.[61]

When Shandong water carriers moved into the Outer City, they transplanted the system of well houses from the Inner City to the Outer City and gradually expanded their business into Chinese communities using their connections with banner people and Chinese officials. Shandong water carriers staked out their distribution territories in the neighborhoods and claimed exclusive rights to the water supply while the individual Shanxi carriers were unable to resist this takeover. Because of their growing monopoly, Shandong water carriers managed to treat public resources such as street wells as if they were their own private assets that were beyond the control of the local communities. This dramatically jeopardized the residents' and the non-Shandong carriers' freedom to fetch water in the Outer City. The worsening situation led to a need for state intervention. As early as 1730, Emperor Yongzheng issued a decree to address this problem: "Water carriers who monopolize wells and the water supply by means of a water distribution territory will be punished. Local officials who learn about a private water den and fail to report it will be penalized as well."[62] Nonetheless, the tendency toward water monopolization was not curtailed. As a consequence, in 1770 the Qianlong emperor again warned:

> The drinking water of residents in Beijing is distributed by water carriers who usually control the water supply of twenty or thirty households. Taking this as an exclusive right, water carriers can even sell this right. Henceforth, the local official should abolish these territories. Water carriers who monopolize the water supply, randomly raise water prices, or take their territories as private property will be accused with the crime of market monopoly.[63]

However, this decree seemed to have little effect on the water carriers in the Outer City. Unlike their counterparts in the Inner City who were rewarded for

60 Wang Fu, *Wangtaichang ji*, vol. 2. (王符撰，王太常集，北京：藝文印書館，卷 2 遺事。)

61 *Qinggaozong shilu*, vol. 96. (清高宗實錄 卷 96，乾隆四年七月乙巳條。)

62 *Qianlong Qingding daqing huidian zeli*, vol. 150, 13–14. (乾隆 欽定大清會典則例 卷 150，都察院六·霸占，第 13–14 頁。)

63 *Guangxu Qingding daqing huidian zeli*, vol. 1039, 7–8. (光緒 欽定大清會典事例，臺北：新文豐出版公司，1976，卷 1039，都察院·五城，第 7–8 頁。)

their loyalty as water carriers, most Shandong water carriers in the Outer City had no commitment to the state. Since the state did not drill the wells, it could impose very limited constraints on the water carriers; meanwhile, the state deliberately neglected water governance. As a result, water distribution territories became, de facto, private property. In the Qianlong period, the state, which assumed no obligation to supply water in the Outer City, had to acknowledge this reality and placed responsibility for water provision in the hands of private carriers. Shandong water carriers thus came to dominate the water supply of the Outer City. In exchange, water carriers were required to pay a well rent charge and a water sales tax to the state.[64] In this way, water carriers parcelled out service areas among themselves and served all of Beijing.

6 The Regime of the Water Carriers

In the Inner City, the Shandong water carriers succeeded in manipulating the power of political patronage and creating a well house system with water distribution territories that adhered to the residential patterns of the banner people. Later, when water carriers transplanted the well house system to the Outer City, they controlled the water supply across the whole city. In their daily business, the well houses and water carriers defined their water regime by compartmentalizing Beijing and they exercised their water power through distribution territories. Each individual water carrier of a well house dominated the water supply in a specific region as an exclusive right. For example, in the Guangxu period, water carrier Yan Quan distributed water to households located in five alleyways.[65]

Beijing thus received its water by way of a water regime in which water carriers, through their water distribution territories, collected and distributed water to householders and created a spatial order in the city that was outside of the

64 Chinese First Historical Archives, No. 21-0301-0040, October 12, 1907. (中國第一歷史檔案館，21-0301-0040，为外城总厅申请豁免民地开井地租事给会计科移文，光绪三十三年十月十二日。)

65 The following alleyways include Shier Hutong (石儿胡同), Huangque Hutong (黄雀胡同), Beihengjie (北横街), Nanwulao Hutong (南五老胡同), and Siniujiao Hutong (死牛角胡同). See Oriental Culture Institute of Tokyo University, *Dongyang wenhua yanjiusuo cang zhongguo tudi wenshu mulu*, 123–131. (濱下武誌等編，東洋文化研究所藏中國土地文書目錄·解說（下），東京：東京大學東洋文化研究所附屬東洋文獻中心，1986 年，第 123–131 頁; The Contract housed in the Niida Collection, Tokyo University, Oriental library, No. 14–335).

order constructed by the armed and bureaucratized men of the Qing state.[66] The privilege of supplying water to the people of a given district would ordinarily belong to one water carrier who had acquired this right through inheritance or purchase. If he did not wish to run the business himself, he would lease the privilege and receive 10 percent of the total income of the district.[67] For instance, water carrier Deng rented out his inherited distribution territory at thirty taels of silver for two years.[68] Thus, water carriers were continually defining and disputing their territories in daily operation. For example, water carriers in Liulichang (琉璃厂) fought for the water distribution rights to a shop and filed a lawsuit in court.[69]

The water control mechanism had less to do with the ownership of water than the choices made concerning its allocation. To a large extent, the water regime maintained its power through exclusive control over the distribution of water rather than water or well ownership. As a result, having a water distribution territory was crucial to a well house, because it meant that each individual well was dispensable; one could always buy water from other wells and then distribute it in one's territory. Well houses with salty water or exhausted wells either purchased water from other well houses, or they rented wells. For example, in the Inner City, some banner elites who were privileged to own wells engaged in the water business. In one case, the bondservant Jilun, registered in the Plain Yellow Banner, subleased a well, four shoulder poles, and two troughs to a Shandong man named Shao in 1767. Later on, Shao's son renewed this contract and hired two water carriers from Shandong to sell water until 1816.[70] Another bannerman named Dechui, under the Plain Red Banner, rented out one well with its distribution territory to the water carrier Wang from 1879 to

66 Alison Dray-Novey, "Spatial Order and Police in Imperial Beijing," *The Journal of Asian Studies*, 52, no. 4 (Nov. 1993): 885–922.

67 Sidney D. Gamble, *Peking: A Social Survey* (New York: George H. Doran Company, 1921), 120.

68 Chinese First Historical Archives, No. 06-01-001-000607-0252, November 23, 1888. (中國第一歷史檔案館，06-01-001-000607-0252，爲訊辦正紅旗宗室孀婦保李氏遺告挑水人董從珠等不準接做水道買賣一案事致宗人府，光緒十四年十一月二十三日。)

69 *Shenbao*, April 7, 1896, p. 1. (《申报》，1896 年 4 月 7 日，第1版。)

70 Chinese First Historical Archives, *Wei shenming zhenghuangqi baoyi yuanren Sichuan pingshanxian Jilun zhinv Erniu hangao tiaoshuiren Shao Kegui deng bugei qi jingzu qianwen yian*, No. 05-08-008-000027-0002, February 7, 1816. (中國第一歷史檔案館，內務府呈稿，爲審明正黃旗包衣原任四川屛山縣濟倫之女二妞喊告挑水人邵可貴等不給其井租錢文一案完結事，05-08-008-000027-0002，嘉慶二十一年二月初七日。)

1884.[71] Water carrier Zhang rented a well from widow Tong from the imperial clan.[72] To make ends meet, the Bureau of Road Works of the Inner City also subleased its three wells to water carriers.[73]

Within an actual well house, above the well mouth was a frame on which pulleys hung, and the ropes over these raised and lowered the bag-like rattan buckets. Two watering troughs often stood nearby. A water carrier emptied water into a tub from which a trough led down to the waiting wheelbarrow that a water seller could trundle off through the lanes.[74] Waste and the overflow water was drained off by a sewer inlet or an open cesspool remodeled from an obsolete shallow well a few feet away. The other trough was for city animals like horses, camels, and donkeys. Drivers needed to pay one cent for watering each animal at that spot;[75] this constituted one source of income for the well house. For instance, water carrier Liu's business declined when the new road bypassed his well and caused a huge loss in clients with animals needing water.[76]

Well houses constantly hired Shandong men as water carriers who were paid by the month, in addition to receiving room and board, shoes, and haircuts.[77] For example, Liu San came to Beijing to carry water at Dongzhi Gate (東直門) in 1775 and saved money in order to take a wife who could attend to his aged widowed mother in his Shandong hometown.[78] The Yi brothers from

---

71 Chinese First Historical Archives, *Muxing zhi shuijing yiyan zhuanzu yu Wang Tingxi ziju*, No. 16-01-003-000155-0035, May 29, 1879. (中國第一歷史檔案館 刑法部档案，穆姓之水井一眼转租与王庭玺字据，16-01-003-000155-0035，光绪五年五月二十九日。)

72 Library of Institute of Modern History at Academia Sinica: *Xiangbaiqi yigu zongshi zaigeng zhiqi hangao Zhang Jisheng bugei jingzu qianwen yian* (1853), Zongrenfu shuotang gao, vol. 10, 555–557. (镶白旗已故宗室载赓之妻喊告张吉盛不给井租钱文一案，《宗人府说堂稿》，咸丰三年正月二十九日 司最字四号，中研院近史所图书馆藏，第 10 册，第 555–557 页。)

73 Chinese First Historical Archives, *Shensong benju shangyue shuijing yuexi qingce shi*, No. 21-0301-0014, May 11, 1907. (中國第一歷史檔案館，申送本局上月水井月息清册事，21-0301-0014，光绪三十三年五月十一日。)

74 "Milk and Water in Peitaiho," *The Missionary Herald*, November 1913, Boston, American Board of Commissioners for Foreign Missions, 509.

75 *Jingdu shuigui, Shenbao*, June 10, 1875. (申報，光緒乙亥六月初十日，京都水貴。)

76 Chinese First Historical Archives, *Shenbao zhuojian dongsanzuomen nei shuijing yuezu shi*, No. 21-0301-0005, February 13, 1907. (中國第一歷史檔案館，申报酌减东三座门内水井月租事，21-0301-0005，光绪三十三年二月十三日。)

77 Sidney D. Gamble, *Peking: A Social Survey* (New York: George H. Doran Company, 1921), 120.

78 Chinese First Historical Archives, *Tiwei jingshi Liu San oushang yiqi Wangshi shensi yizhui jiaojianhou shi*, No. 02-01-07-1496-006, September 26, 1777. (中國第一歷史檔案館 內

Shandong ran a water business at Youan Gate.[79] Literate Shandong men might work as accounting clerks in a well house.[80] The well house and its native-place sentiment enhanced the solidarity and cohesion of water carriers. When dealing with disputations, a well house would react collectively. For example, when one private carriage driver quarreled with a water carrier, he later went with weapon in hand to the well house to seek retaliation. The irritated well owner ordered all his carriers to strike the driver with shoulder poles. Unfortunately, the carriers beat the outnumbered driver to death.[81]

Water from wells was conveyed to the door in a one-wheeled cart bearing two tubs on each side. One water wheelbarrow load contained approximately twelve buckets of water, and the water carriers would sell this water to householders. The carrier needed nothing but the shrill squeak of his wheel to announce his coming and bring his patrons to their doors. When delivering to a house, the water was emptied into two wooden pails by taking out the wooden plugs in the two holes in the sides of the tanks. A hardy carrier would push a cart holding a large barrel of water drawn from a well and work every day from early morning to late evening, in scorching summer and frigid winter (except for Chinese New Year). Working conditions were especially strenuous during the summer rainy season, when the unpaved streets and lanes were almost impassable.

The well houses were confronted with various challenges in their daily operations. First, the water carriers were prone to traffic accidents as they pushed their heavy wheelbarrows filled with water in the streets. One working water carrier was killed on the spot by a speeding carriage at Xisi (西四) and the hit-and-run driver was caught later.[82] Another water carrier's water cart ran over a cart loaded with sesame oil and emptied it. The well house had to make compensation for this high-priced loss.[83] Second, the accessibility of the well houses in the streets tended to invite mishaps. For example, a bankrupted gambler sneaked into the Dragon King temple of a well house one night and hanged himself. The local government requested that the well house buy a coffin for

阁刑科题本，02-01-07-1496-006，题为京师刘三殴伤伊妻王氏身死议准绞监候事，乾隆四十二年九月二十六日。)

79 Chinese First Historical Archives, *Zhongguo diyi lishi danganguan*, No. 02-01-07-06663-003. (中國第一歷史檔案館藏：內閣刑科題本土地債務，貪汙，違禁類02-01-07-06663-003，題爲會審京師人夏三因碰倒水車爭鬧毆傷尹二身死一案依律擬絞監候請旨事 乾隆三十六年七月二十五日。)

80 *Shenbao*, July 9, 1889. (《申报》，1889 年7月9日，第 2 版。)

81 *Shenbao*, August 16, 1886. (《申报》，1886 年8月16日，第 2 版。)

82 *Shenbao*, May 28, 1886. (《申报》，1886 年 5 月 28 日，第 1，2 版。)

83 *Shenbao*, November 4. (《申报》，1890 年 11 月 4 日，第 1，2 版。)

the deceased because it had failed to lock the door of the temple.[84] The owner of another well house had two troughs to store water, and a man with mental confusion drown himself in one of the troughs one night.[85] Finally, a well house could also fall victim to violence and the militancy of other occupations. For example, a bricklayer extracted water at one well house for construction, but he lost his jar in the well, then had an argument with the water carriers. The raging bricklayer called upon his construction team workers to attack the well house. With rods in hand, they injured five carriers.[86]

Within the water carriers' regime, the subjects of the Inner City and the Outer City were treated differently. Water carriers and *erxingzi* water they distributed from the state-owned wells became a symbol of banner identity in the Inner City. The reason behind this is that the common element in the Beijing banner people's diet was old tribute rice (老米) that was brought from the south Yangtze River region through the Grand Canal. Before reaching the hands of the banner people, this rice was stored in granaries for two or more years due to the long-distance transportation and distribution system. When cooked with *erxingzi* water, this stale rice yielded a special flavor that was favored by the Manchu and gradually shaped the tastes of the banner people.[87] *Erxingzi* water gradually became one of the symbols of the banner people and persisted into the late Qing. After two hundred years of residence in the Inner City, the glory and wealth of the banner people had vanished with the decline of Qing Empire, and impoverished banner families could not even afford drinking water. For example, Lao She (1899–1966), a noted writer and also a Manchu registered under the Plain Red Banner, was born and raised in the northwest part of the Inner City. During his childhood his family could barely afford the water distributed by water carriers, but they still used this service because water carriers and the *erxingzi* water they supplied were symbols of their Manchu identity and social class.[88]

When banner families became penniless, the water carriers would aggressively argue with them about outstanding water bills. As a consequence,

84 *Shenbao*, December 1. (《申报》，1890 年 12 月 1 日，第 1，2 版。).

85 Chinese First Historical Archives, *Wei chahe Li Shun zai Long Yongzhou shuicaonei beiyan shensi an*, No. 16-01-017-000036-0046, February 1904. (中國第一歷史檔案館藏：刑法部档案，档案号：16-01-017-000036-0046，为查核李顺在龙永舟等水槽内被淹身死一案事，光绪三十年二月。)

86 *Shenbao*, October 5, 1895. (《申报》，1895 年 10 月 5 日，第1，2版。).

87 Jin Shoushen, *Life in Old Beijing* (Beijing: Beijing chubanshe, 1989), 212–213. (金受申，老北京的生活，北京：北京出版社，1989 年, 第 212–213 頁。)

88 Lao She 老舍, *Zheng hongqi xia* 正紅旂下 [The Red Banner] (Beijing: Renming wenxue chubanshe, 1980).

conflicts between water carriers and bannermen were not rare. Nevertheless, banner society was able check the power of water carriers. First, the wells were owned by the state in the Inner City, and water carriers were subject to the state. Even in the final decade of the Qing dynasty, the wells were still in the control of the state. Second, the deeply bureaucratized and paternalistic banner system itself wielded tremendous power and played a significant role in counterbalancing the power of water carriers. In the Qianlong period, one local banner officer warned water carriers in his jurisdiction to control their behavior.[89] Last but not least, many of the banner people were military men, which deterred the water carriers' recklessness. For example, during the Qianlong period, the water carrier Chi Si abused Bo Lin, a Bordered Yellow bannerman, over water bills; Bo was so furious that he killed Chi.[90] Another example is Ge Tongge, the chief of a company in the Bordered White Banner, who killed his water carrier Zhou Ba with an axe because Ge suspected that Zhou was having an affair with his concubine.[91] To a large degree, in the Qing period, water carriers in the Inner City were compliant and the water supply was more orderly when compared to the situation in the Outer City.

Unlike their counterparts in the Inner City, the neighborhoods of the Outer City were composed of migrants, including clerks, shoppers, craftsmen, merchants, and students from the empire. These temporary tenant communities lacked solidarity and were exploited by Shandong water carriers, who faced no opposition. The water carriers would take advantage of every opportunity to make money. For instance, in June 1875, water carriers doubled the price of water delivery because of drought. A well-known Chinese official admitted that his supply of water from the springs cost him as much as 300 *yuan* a month, which is about the cost of an annual water bill for an ordinary household.[92] In fact, according to one newspaper, the high price was not caused by water scarcity; rather, on the pretext of drought, the water carriers tyrannized people because of the monopoly they had created through the system of water distribution territories.[93]

---

89 Song Yun, *Baier laoren yu*, vol. 4. (松筠，百二老人語，第4冊坎部卷2旗員事9條，首都圖書館古籍部特藏。)

90 (Zhongguo dili yili danganguan, no. 02-01-07-07564-004，題爲會審京師民栢林因被索討水錢爭毆傷斃池四一案依律擬絞監候請旨事，中央研究院歷史語言研究所 內閣大庫檔案 乾隆四十六年閏五月二十九日。)

91 (Zhongyang yanjiuyuan shiyusuo, no. 221348-001, 中央研究院歷史語言研究所 內閣大庫檔案 乾隆 47 年 11 月 10 日 刑部爲格通額毆死周八由 登陸號：221348-001。)

92 Juliet Bredon, *Peking: A Historical and Intimate Description of Its Chief Places of Interest* (Shanghai: Kelly & Walsh, 1920), 61.

93 *Jingdu shuigui, Shenbao*, June 10, 1875. (申報，光緒乙亥六月初十日，京都水貴。)

TABLE 1 Annual water expenses of the Li Ciming household (—represents data unavailable)

Years	Sweet water payout (dollar)	Brackish water payout
1871	30	40
1874	35	60
1877	—	40
1878	280	43
1881	180	340
1882	380	—
1883	300	—
1884	278	—
1887	442	—

SOURCE: ZHANG DECHANG, *THE LIFE OF AN OFFICIAL IN BEIJING IN LATE QING* (HONG KONG: CHINESE UNIVERSITY OF HONG KONG PRESS, 1970), 68, 150.

Because of the limited supply options, the residents of the Outer City had to rely on the water sold by the carriers, but these migrants complained that Beijing's water was the most unpalatable of drinking waters, whether bitter water or sweet water. Even worse, water carriers monopolized and sold it at a high price, so residents in Beijing were more afraid of the price of water than the price of rice.[94] When using carriers' water, residents would purify it by the addition of alum.[95] The following table shows the water bill of an official family in the Outer City in late Qing Beijing. In this table we can see that both sweet and brackish water were commodities, and even an official family could not afford to use sweet water exclusively.[96]

The price of water supplied by carriers was so unbearably high that water carriers became satirical objects in folk literature. One Beijing ballad recorded by missionaries in the late nineteenth century mocked the water carriers because of their bad pronunciation and awkward manners. This song, sung

94 *Yanjing zaji* (Beijing: Beijing Guji Chubanshe, 1986), 133. (闕名：燕京雜記 北京：北京古籍出版社，1986 年。)

95 Charles Alexander Gordon, *Remarks on Certain Points Relating to Public Health in China*, vol. 19 (London: William Clowes and Sons Limited, n.d.), 210.

96 Zhang Dechang, *The Life of an Official in Beijing in Late Qing* (Hong Kong: Chinese University of Hong Kong Press, 1970), 68, 150.

by boys behind the water carriers' backs, defined the carriers as turtles. The word "turtle" in China was used for one of the direst insults, as this animal was phantastically charged with an unnatural crime.

> Water carrying fellow,
> Hear what I say.
> On the bank of the south river is your hole.
> When the weather is fine, you dry out your shell;
> When the weather is bad, then you draw in your neck.[97]

Nearly all residents felt the pressure of the rising water prices, and consequently they had to conserve water. Sweet water was remarkably expensive, and housewives usually diluted it with bitter water.[98] Each house was supplied with two jars, one for sweet water and the other for bitter water. While conserving their water supply, residents also spared no effort to collect rainwater in basins and all manner of vessels for domestic use and tea infusions. Before collecting the water, sufficient time had to be allowed for the rain to first thoroughly cleanse the roof tiles, which retained a large amount of dust and organic and calcareous matter. Rainwater collection was not limited to the Chinese but also included foreign families who dug large cisterns into which the water from their roofs drained. A plentiful supply of clean water was thus to be had throughout the year for washing and watering their flower gardens in spring and autumn.[99]

7 Conclusion

In multiethnic Beijing with its notoriously salty well water, the priority of the state was to secure safe and drinkable water for the banner people. The water carrier systems were created by the state as a means to municipalize drinking water, so that the state had direct control of supply and indirect control of distribution. Patronized by the state, Shandong water carriers invented the well house system and water distribution territories according to the divisions

97 Baron Guido Vitale, *Pekinese Rhymes: First Collected and Edited with Notes and Translation* (Peking: Pei-T'ang Press, 1896), 7–8.

98 Chu Weikai, "Yanjing zayong," in Lei Mengshui, ed., *Beijing fengsu zayong* (Beijing: Beijing Guji chubanshe, 1982), 52. (褚維塏：燕京雜詠，雷夢水，北京風俗雜詠 北京：古籍出版社，1982 年。)

99 John Dudgeon, *Diet, Dress and Dwellings of Chinese in Relation to Health* (London: William Clowes and Sons, Limited, 1884), 296–97.

in the residential quarters of the banner people, and they later extended this system to the entire city. Access to water was secured through complex institutional arrangements based on geographical territories. The state water carriers in the Inner City and the private water carriers in the Outer City also demonstrate the profound disparity between banner and non-banner people. While the power of state water carriers was counterbalanced by the state and by the banner system in the Inner City, the private water carriers in the Outer City, who faced less opposition by the tenant communities, were more tyrannical.

The water regime as urban infrastructure was built on water distribution territories and native-place sentiment, which led to exclusive, solidified, and territorialized power. Water carriers secured a supply of water for the city and created a spatial order that was outside of the order enforced by the bureaucratized men of the state. Within their regime, each well house enjoyed its own local monopoly in its service area, and city dwellers lost the ability to switch carriers when their service lagged. In this way, water carriers shaped Beijing residents' habits of consuming water and wove their delivery routines into the fabric of daily life. Moreover, the water regime enhanced the water carriers' ability to navigate modernity after 1900.

CHAPTER 2

Foreign Well Water, 1900–1909

1 Drinking Water and the Boxer Rebellion

Following the Second Opium War (1856–1860), the Qing government was forced to grant Westerners and missionaries a permanent presence in Beijing.[1] They built the legations quarter and churches in the city and converted thousands of Chinese to Christianity. In the wake of the humiliation of the Opium War, Chinese frustration at unbridled foreign aggression finally found expression in the Boxers, who orchestrated brutal attacks on privileged foreigners and their Chinese converts with the intent of driving them from China. This anti-foreigner sentiment was shared by a range of ordinary Chinese people. While scholars tend to argue that most Boxers were peasants, it has been proven that different professions including businessmen, officials, scholars, and even prostitutes, became involved in the Boxer rebellion.[2] The breadth and depth of sentiment against foreigners shaped politics in the last decade of the Qing dynasty.

In June of 1900, the Boxers from north rural China, mainly from Shandong Province, with their extreme anti-Christian and anti-foreigner sentiments, swarmed into Beijing to hunt down foreigners and any Chinese they considered tainted by foreign contact. The Boxer mobs, supported by the Qing government, looked upon their leaders as deities with magical powers and would set up altars to them. After an altar was established, Boxer fighters stationed themselves in the streets and went house to house in that neighborhood to request pancakes. Even impoverished families, both willingly and unwillingly, prepared and dedicated pancakes, which were piled up like hills around each altar, to the fighters.[3] It is not surprising that the water carriers sympathized with the Boxers, since both groups hailed from Shandong Province. While city dwellers contributed food, water carriers were enlisted to supply free drinking water. Since June and July were the scorching season in Beijing, the Boxers demanded large amounts of water to quench their thirst.

1 The Second Opium War is also known as the Second Anglo-Chinese War, which arose from the failure of the Treaty of Nanjing (the aftermath of the First Opium War, 1839–1842) to satisfy British goals of advancing trade and diplomatic relations with China.

2 Chai E, "Gengxin jishi," in Jian Bozan, ed., *Yihetuan* (Shanghai: Shanghai renmin chubanshe, 1957). (柴萼：庚辛紀事，翦伯贊等編 義和團 上海：上海人民出版社，1957 年。)

3 Chai E, "Gengxin jishi," in Jian Bozan, ed., *Yihetuan* (Shanghai: Shanghai renmin chubanshe, 1957). (柴萼：庚辛紀事，翦伯贊等編 義和團 上海：上海人民出版社，1957 年。)

 | DOI:10.1163/9789004720800_004

The water carriers followed Boxers' orders to distribute water to each altar, as depicted in this bamboo twig poem (竹枝詞):

> The water wheelbarrow rocked the street, with a dragon flag;
> A scoop of water can slake thirst, since Boxers don't drink tea.[4]

Because the Boxers were thought to possess spirits, the water wheelbarrow that supplied them needed to be distinguished from other wheelbarrows, so it was decorated with a flag bearing the characters "Divine Boxer" (神拳). Furthermore, the Boxers drank only water; tea was strictly forbidden because it was believed to jeopardize their magical invulnerability to foreign weapons. The rest of the water carriers, who delivered water for ordinary households, were asked to show their respect and to identify with the Boxers by wearing yellow or red scarves on their heads, as the Boxers did.[5]

Although the legations and Christian communities had relied on water carriers for potable water before the rebellion, it was not long until the water carriers refused to deliver water to foreigners and Christian families within their distribution territories. Soldiers had to escort Chinese Christians to the wells and stand guard while they drew water. At the beginning of the legation siege in June, the question of access to a water supply and food confronted the foreigners and Chinese Christians who sought refuge in the legations quarter. As one British woman wrote in her diary: "The second day we had been in the Legation, one of the servants came in and said: 'What shall we do for water? They have locked the well.' My heart sank, for I feared the reason of a locked well was scarcity of water."[6] Lest a water famine ensue, a water committee was appointed to take charge of the five wells in the British Legation and its duty was to measure the depth of water each day and regulate its consumption. Fortunately, these five wells in the compound supplied abundant clear drinking water.[7]

Water seemed to be used as a kind of weapon by the Boxers to subdue foreigners and their followers. Water carriers also had a hand in the Boxer movement, but there is no evidence indicating that Shandong water carriers

---

4 Fu Nongshi, "Dumen jibian baiyong," in Lu Gong, ed., *Qingdai Beijing zhuzhici* (Beijing: Beijing chubanshe, 1962), 255–256. (復儂氏: 都門紀變百詠，路工編選，清代北京竹枝詞（十三種）北京：北京出版社，1962年，255–256。)

5 Lin Yutang, *Moment in Peking* (New York: John Day Co., 1939), chap. 1.

6 A. H. Mateer, *Siege Days: Personal Experiences of American Women and Children during the Peking Siege* (New York: Fleming H. Revell Company, 1903), 92, 197.

7 Report of Events in Peking from the 20th June to the 14th August, 1900, Further Correspondence respecting Events at Peking [In continuation of "China No. 4 (1900)"], Great Britain, Foreign Office Correspondence, vol. 4 (1900–01), June 21, p. 3.

manipulated their familiarity with the neighborhoods to collaborate with their fellow Shandong townsmen to hunt down and loot households during the upheaval, as their Shanxi predecessors had done during the Ming-Qing transition in 1644. Nevertheless, the outbreak of the Boxer Rebellion interrupted the order of water regime since many inhabitants escaped from the city, causing well houses to run at a loss. For example, in 1900, water carrier Wu even asked his lessor to extend his contract for one more year as compensation for the deficits arising from the Rebellion.[8] In any event, the presence of the Boxers in Beijing's streets only lasted about two months. But while the impact imposed by the Boxers themselves on water and its carriers was provisional, the influence exerted on water governance by the new international coalitions established in the city in the aftermath of the Boxer Rebellion were enduring.

When the international legations in Beijing where foreigners and Chinese Christians sought refuge came under attack by Boxer rebels in June 1900, the eight nations of Japan, Russia, Britain, France, the United States, Germany, Italy, and Austria set up the Eight-Nation Alliance, an international military coalition, to lift the siege. During an expedition to Beijing, the allied armies had learned of the scarcity of drinkable water.[9] Following the lifting of the siege, Beijing was occupied by seven nations: the Inner City was divided into Japanese, Italian, Spanish, and Russian zones, while the Outer City was partitioned into British, American, and German zones.[10] Upon settling in Beijing in August, an international expeditionary force under the command of General Waldersee of Germany soon discovered that drinking water was a pressing problem because of the city's notoriously bitter water. Initially the allied forces attempted to fetch water from the Jade Spring, the source of imperial drinking water. However, this turned out to be impractical owing to the long distance and tremendous demand. Finally, the Germans determined to drill wells to secure a supply of drinking water, and used the Western method of machine-aided percussion drilling to sink a deep well on Yingtao Skewed Street (櫻桃斜街)

8 Chinese First Historical Archives, *Wei heyi neiwufu yuanwailang achanga tanpi jiading qiangba shuijing an*, No. 16-01-005-000023-0081, July 1903. (中國第一歷史檔案館：为核议内务府员外郎阿昌阿祖庇家丁强霸水井奏请解任听候审办一案事等，档案号：16-01-005-000023-0081，光绪二十九年七月。)

9 "Shanghai," *North China Herald*, July 25, 1900, 201. Also, Carl R. Darnall, who served in the US Army as an assistant surgeon during the Boxer Rebellion, later discovered that liquid chlorine could purify water and developed an advanced water filter (Eboch, M. M., *Medicine on the Battlefield*, Essential Library, 2016, 53–55).

10 As an ally of Germany, Austria did not participate in the occupation. See Michael Hunt, "The Forgotten Occupation: Peking, 1900–1901," *Pacific Historical Review* 48 (November 1979): 501–529.

in the Outer City.[11] This well-drilling technology enabled them to deepen Beijing's wells and improve the water quality, which would impact water's relationships to the city and water carriers.

Japan's involvement in Beijing's drinking water supply began in 1900 as well. The Japanese dispatched more than 20,000 troops to China, accounting for two-fifths of the total number of allied forces used to overcome the Boxer siege. After settling in the Inner City, the Japanese military needed to secure reliable drinking water sources for its troops and drilled a well on East Forty-Second Street (東四十二條) using Japanese technology, which differed from Germany's well-drilling technology in terms of both its sources of power and its construction materials. Japanese well drilling relied on manpower while the Germans used steam engines, and the Japanese cased the well's borehole with bamboo pipe while the Germans used iron pipe. Unfortunately for the Germans, both steam engines and iron pipe were unavailable in preindustrial Beijing. Thus, only the Japanese well-drilling technology proved economically viable and it flourished in Beijing while the German occupation forces sold their well the following year because of the prohibitive cost of further well construction. Soon after, the Japanese established well-drilling enterprises to popularize their well-boring methods in Beijing. The Boxer episode brought the Japanese into close relations with the Chinese, and thus China's neighbor became interested in developing China in many ways including infrastructure, education, and military readiness, with the sinking of wells as one of the most promising of these. This repercussion of the Boxer Rebellion led to the dramatic transformation of well drilling and the ensuing urban water reform. The Boxer Rebellion was, in this sense, a watershed in the history of Beijing's drinking water.

2 Japanese Well-Drilling in Beijing

Well-drilling technology played an important role in procuring drinking water in preindustrial cities, especially in China, until the early twentieth century.[12]

---

11 Unidentified author, *The General History of Foreign Wells in Beijing, The Trivial Record of Beijing's History*. (出版者不詳，燕京今古瑣聞錄，民國, 北京市洋井之略歷。)

12 Both the Inner City and Outer City had their own well-drilling artisans. For example, when the well-drilling technician Hu Yiming who served under the Plain Yellow Banner died of illness, his brother Hu Yide, also a well-drilling expert, was willing to substitute for him after an interview. See Chinese First Historical Archives, *Wei zisong zhaomu jingjiang Hu Yide bing yinjie cunjuanshi zhi neiwufu*, No. 05-013-002-000392-0080, April 19, 1756. (中國第一歷史檔案館：为咨送招募井匠胡义德并印结存卷事致内务府，档案号：05-013-002-000392-0080，乾隆二十一年四月十九日。) Liu Ba, who lived

The principal sources of drinking water in Beijing were hand-dug wells, which were usually thirty feet in depth. However, the plain upon which Beijing sits consists of layers of loess, clay, sand, and gravel, with the productive good-water zone at a depth of around 200 feet (see table 2).[13] The fact that wells were dug by hand limited access to this deeper and better-tasting groundwater, so the simplest and most efficient way to improve Beijing's drinking water was to deepen its wells to break the bounds of its previous environmental limits.

China did have a drilling technology for boring salt wells, but this method wasn't suitable for water wells in Beijing, thereby creating an opening for the Japanese method. Well boring did not simply consist of sinking a hole in the ground; rather, it was a complex process that included determining the drilling site, test boring, sinking, casing the well, and maintenance.[14] The application of the percussion method in 1041 of the Song dynasty revolutionized a Chinese technique in Sichuan Province for drilling salt-mine boreholes from which brine was ingeniously raised.[15] This method developed by China used frameworks constructed of bamboo that allowed a crushing tool to be raised and dropped, and produced wells drilled to the depth of one thousand feet.[16]

---

on Ox Street in the Outer City during the Kangxi period, was a professional well-drilling craftsman and residents who intended to dig wells would contract him. See *Beijing niujie zhishu-Gangzhi* (Beijing: Beijing chubanshe, 1990), 18. (北京市政協文史資料研究委員會等編，北京牛街誌書-岡誌，北京：北京出版社，1990年，第18页。)

13 Bailey Willis, *Research in China*, vol. 1, pt. 1 (Washington: Carnegie Institution of Washington, 1907), 196.

14 The history of well-drilling technology in China has not been thoroughly investigated. Well-drilling technology has occasionally been addressed in studies of the salt industry in China. See Liu Chunyuan, "Woguo songdai jingyan zuanzhao gongyi de zhongyao gexin-Sichuan zhuotongjing," *Wenwu* 23 (December 1977): 66–72. For the most recent research, see Pan Jixing, "Zhongguo shenjing zuantan jishu de qiyuan, fazhan he xichuan," *Yanyeshi yanjiu* 23 (October 2009): 3–33.

15 In *Science and Civilization in China*, Joseph Needham briefly mentions that in the early Han period (206 BCE–9 CE) the Chinese had fully developed the technique of drilling boreholes two thousand feet deep; see Joseph Needham, *Science and Civilization in China*, vol. 4, pt. 2 (Cambridge: Cambridge University Press, 1965). However, this argument is incorrect. In fact, the application of deep drilling in wells occurred around 1041 CE in the Song Dynasty instead of in the

16 Madeleine Zelin, *The Merchants of Zigong: Industrial Entrepreneurship in Early Modern China* (New York: Columbia University Press, 2005), 14–16. *Tiangong kaiwu*, (天工開物, 1637), a book on various industries and labor-saving machinery, gave an accurate description of the processes involved in boring salt wells. The salt-well drilling in Sichuan was also discussed by American scholars in the twentieth century; see Crawford Wallace, "The Salt Industry of Tzeliutsing," *China Journal of Science and Art* (Shanghai) 4 (1926): 174; and

TABLE 2 The layers of the Beijing Plain from a condensed well-log of the deep well at the Belgian Legation of Beijing in 1903.

Well-Log	Component	Depth (feet)
	Ground level	0
	Soil and debris	33
	Sand water-table	36
	Sand	39.5
	Coarse sand and gravel	141.5
	Sand	144
	Coarse sand and gravel	162
	Clay	164
	Sand	168
	Clay	172
	Sand	174
	Clay	177

SOURCE: GEORGE B. BARBOUR, "DEEP WELLS IN THE PEKING AREA," *BULLETIN OF THE GEOLOGICAL SOCIETY OF CHINA* 3, NO. 2(1924): 136.

However, one of the conditions for popularizing a technology is its lower expense of output; otherwise there will be no market for it.[17] The process of salt-well construction was both time consuming and economically infeasible; it generally required years to bore a salt well of medium depth. One author stated that one day's labor might result in four to five inches of progress, and it usually took between four and ten years to reach brine.[18] Accordingly, this well-drilling method was confined to salt mining in the remote southwest region of China, and hand-dug wells for water were prevalent in north China prior to 1900.[19]

Archer, M. T. "Drilling in Tzu-Liu-Ching, China," in *History of Petroleum Engineering*. Dallas, TX: American Petroleum Institute, 1961: 146–52.

17 "The Making of a New China," *The North China Herald*, October 30, 1903, 894.

18 Li, Jung, "An Account of the Salt Industry at Tzu-liu-ching," translated by Lien-che Tu Fang, *Isis* 39, no. 4 (1948): 228–34.

19 Ironically, the method of salt-well drilling identified as *Chinesische Seilbohren*, or *Methode der Chinese*, was introduced to Europe by Jesuits and impacted well drilling in Germany and France before the advent of machinery drilling in the middle of the nineteenth century. For example, François Imbert, (1800–1870), a French Jesuit in Sichuan province

The well-drilling method the Japanese used in Beijing was named the Kazusa System, or Kazusabori (上総掘).[20] Invented in the 1870s in the Kazusa area (currently Chiba Prefecture, 千葉縣) of Japan, this system uses a rig constructed of wooden derricks and cable tool drilling, which makes it possible to efficiently dig wells to a depth of several hundred feet using only a few people. The Kazusabori method spread throughout Japan in the late nineteenth century and was even used to drill hot springs and bore oil fields.[21] Although this Japanese well-drilling method has some similarity to the Chinese system of boring employed in Sichuan, the connection between them is unverified. Some scholars believe the Japanese method owes its origin to Sichuan based on the fact that the Japanese had historically been keen borrowers of Chinese ideas and notions.[22]

The Kazusa System of well boring utilized the following steps: (1) select and acquire a drilling site; (2) erect a platform on which a lever is built, with an iron chisel suspended from a cable attached to the lever; (3) manually raise and drop the lever to cause the heavy drill to fall, breaking through the clay and rock bed, with one laborer guiding the drill cable as it descends from the scaffold to make sure that when the drill falls it creates a cavity that is both round and straight; (4) pull up the iron pipe, now filled with excavated earth and sand. During the process, three precautions had to be strictly followed: (1) the interior of the drilling hole had to be piped with bamboo immediately after lifting the rig to prevent collapse; (2) the mud in the hole of the rig had to be examined to predict the quality of the water; and (3) construction had to be a

around 1825, conducted a detailed investigation of salt well drilling and sent the findings back home. François Imbert, "Lettre de M. Imbert." The Chinese percussion method and rope bore inspired European well drilling; see Fromann, C. W. *Die Bohr-Methode der Chinesen oder das Seilbohren*. Coblenz: 1835; and Jobard, J. B. "Uber das Brunnenbohren mit dem Seil," Polytechisches Journal (Stuttgart) 105 (1847): 14–24.

20 F. J. Norman, "Boring in Japan," *Transactions of the Institution of Mining Engineers* 23, (1903): 685–699.

21 Hiroki Takamura, "The Techniques and Tools of the Kazusabori Well-boring," *Journal of Earth Environment Research* 6 (2004): 37–49.

22 See F. J. Norman, "Boring in Japan," *Transactions of the Institution of Mining Engineers* 23 (1903): 685–99. Other scholars speculate that the well-drilling method devised by the Chinese was introduced in Japan sometime after the beginning of the Sui Dynasty in 618 CE, but no documentary evidence remains; see Allied Powers, Supreme Commander, *Preliminary Study*, Natural Resources Section, issues 49–64, 1951.

nonstop process, otherwise the iron rig would become stuck and the project would come to a halt.[23]

The Kazusa System was innovative because the apparatus attached to the tip of the drilling rod was made of a piece of bamboo that could be added to line the wells. The system also used muddy water and clay to prevent the well from collapsing.[24] By using bamboo tubes tipped with steel points, Japanese drillers could drive through hundreds of feet of surface earth, sand, gravel, and other strata to the subterranean levels of pure water, which then rose near enough to the surface to be pumped up without difficulty. Hence by digging a well of several feet in diameter and to a depth of three or four feet below the water table, then driving down to the deep stratum bearing deep water, the open well could serve as a sump or reservoir from which water could be drawn with buckets. Compared to hand-dug wells that were limited to thirty feet in depth, these wells could be dug as deep as seven hundred feet. At such a depth, the new wells not only had access to high-quality water but were free from surface pollution and infiltration. This method thus significantly improved water quality.

Compared to German and Chinese well-digging methods, the biggest advantage of the Kazusa System of well boring was its inexpensive cost and simplicity. The entire outfit used in a Japanese drilling operation required an outlay of less than 96 yuan (about 40 US dollars), whereas the necessary mechanical plant obtained from America or Europe was around 3,600 yuan (1,500 US dollars). The other advantage the Kazusa System had for reducing cost was its use of massive bamboo piping.[25] Labor for the Kazusa System generally consisted of five men who filled and let down the tools into the borehole, as well as one skilled man to supervise the well-sinkers. The supervisor sat over the well and directed operations. There was also a boy whose job was to run messages and pass tools as required. In Japan, well construction usually took ten days to a fortnight to complete.[26]

---

23 Hu Zhang, *Dongguo zaojing fa* [The method of Japanese well-drilling], published by Agricultural Society, Guangxu Period (1874–1908), rare book housed at Nanjing Library, China. (胡璋，東國鑿井法，農學會刊行，光緒年間，南京圖書館古籍部藏). Also, see F. J. Norman, "Boring in Japan," *Transactions of the Institution of Mining Engineers* 23, (1903): 685–699.

24 Yasuo Sakura, "History of Hydrogeology in Japan," in Nicholas Howden and John Mather, ed., *History of Hydrogeology*, (CRC Press, 2013), 136.

25 "Kazusa System of Deep Boring for Water," *The Japan Weekly Mail*, June 14, 1902, 645.

26 House of Commons and Command, Parliamentary Papers, vol. 66, Indian Irrigation Commission, Mr. F. J. Norman, Calcutta, Agriculture, 15th December 1902, 276–278.

FIGURE 4 Boring plant and reel
SOURCE: F. J. NORMAN, "BORING IN JAPAN," *TRANSACTIONS OF THE INSTITUTION OF MINING ENGINEERS 23* (1903): 698.

The efficiency of Japanese drilling technology attracted Chinese attention even before the Boxer Rebellion. In the wake of the First Sino-Japanese War in 1894, more Chinese officials and scholars traveled to Japan where they were impressed by Japanese wells and drilling technology during their tours.[27] After 1900, the Japanese, as an occupying force, lifted the Manchu state's prohibition on well drilling in Beijing and Japanese well drilling gradually became popular in the city. By the end of 1909, at least four Japanese well-drilling enterprises had been established in Beijing: Ito Well Drilling Company (伊藤鑿井局), Nisshin Well Drilling Company (日清鑿井局), Nihon Well Drilling Company (日本鑿井局), and Hasegawa Well Drilling Company (長谷川鑿井局). The Hasegawa Well Drilling Company, located at the Lamb Hutong (羊肉胡同) of Dongdan (東單), provided comprehensive services including sinking, repairing, dredging, and maintaining wells.[28] These Japanese well-drilling companies were usually staffed by small Japanese crews. For example, the Nihon Well Drilling Company, located at the South Donkey Market (南驢市) of Xidan (西單), had only four regular employees.[29] However, its advertisement presents a picture of efficiency:

> The Nihon Well-Drilling Company has specialized in well drilling for a few years in Beijing, and all water from wells it has sunk is sweet and pure with no exceptions. Accordingly, it has a lion's share of the market in Beijing's well drilling, and it also contracts and improves defective wells drilled by other companies, for instance, the well at the Commercial Museum.[30]

These Japanese companies were so proficient and professional in sinking wells, lining their interiors, and investigating the sand beds of wells that the water withdrawn from the wells they dug was far superior to the water drawn from Chinese hand-dug wells. The Japanese well drillers opened a new era of well drilling in Beijing.[31] Moreover, the appeal of Japanese well drilling was not limited to Beijing. In April 1902, the governor of Shanxi Province directly

---

27 For instance, Zhang Jian (1853–1926), a well-known entrepreneur and politician, made a special trip to observe the process of well drilling and produced a meticulous description when visiting Japan in May 1903. (*Complete Works of Zhang Jian*, 張謇全集 南京：江蘇古籍出版社，1994 年，第六卷，1903/5/25。)

28 "Japanese Well-Drilling Advertise," *Shuntien Times*, October 1, 1901.

29 Hattori Unokichi 服部宇之吉, et al., *Pekin shi* (*Beijing Gazetteer of Late Qing Period*), Tokyo: Hakubunkan, 1908; translated by Zhang Zongping, *Qing mo Beijing zhi ziliao* 清末北京志資料 (Beijing: Beijing yanshan chubanshe, 1994).

30 "Nihon Well-Drilling Company," *Shuntien Times*, September 5, 1908.

31 Chi, Zehui, ed., *Beipingshi gongshangye gaikuang* (Beiping: Beiping Social Bureau, 1932), 379. (池澤匯等編，北平市工商業概況 北平：北平市社會局發行，1932 年，第 379 頁。)

recruited Japanese craftsmen to instruct the farmers of Shanxi on boring wells, and the necessary drilling outfit was also purchased from Japan.[32] The Nanjing municipal government also hired three Japanese well experts.[33] Even foreign communities in Zhifu (Chefoo) of Shandong Province and Beidaihe (Peitaiho) of Zhili Province contracted Japanese well drillers to secure an abundant supply of pure water.[34]

Japanese well drilling revolutionized the water sources in Beijing, and the water modernization arising from this new technology thus greatly challenged the water regime both in the Inner City and the Outer City. In the Inner City, the state patronized well drilling by establishing the Institute of Well Drilling to retain its control over drinking water; while in the Outer City, merchants and water carriers became the patrons of the new technology. Unlike the state-sponsored well drilling in the Inner City, which was designed to protect the state's sovereignty over drinking water, the water merchants and carriers transplanted Japanese well drilling into the Outer City in pursuit of profits. If the water regime in the Inner City was prompted by the state, the water regime in the Outer City was more oriented by the market.

3 Foreign Wells in the Inner City

Once settled in the Inner City of Beijing in 1644, the Manchu state nationalized all the wells in the streets of the Inner City to heighten the security of drinking water. It further imposed a strict restriction on well drilling with the aim of consolidating the state's monopoly on drinking water. Residents in the Inner City were forbidden to dig wells without permission from the throne. Even though elite families had the privilege of being able to dig private wells, they were not allowed to share their wells with the community. For example, one imperial clansman drilled a well in the open ground outside his door and his neighbors also drew their drinking water from this well. However, a local administrator shut down the well and penalized the owner because he had failed to secure official approval. The owner argued that the well was sunk on private property and no official permission was needed. After the owner's

32 "Japanese in Chinese Government Employ," *The North China Herald*, April 23, 1902.

33 "Zaojing gongshi dihu," *Dalubao* 6 (1905): 5.

34 *North China Herald*, June 11, 1903, and *North China Herald*, Aug 15, 1908, Rocky Point Association (Peitaiho) Annual Report.

repeated petitions, the well was reopened, but only for private use.[35] The prohibition of well drilling was implemented very effectively until the outbreak of the Boxer Rebellion.

In August 1900, the Qing Court was forced to flee Beijing under the pressure of the Eight-Nation Alliance. Following roughly eighteen months of exile, the Court finally returned to Beijing in 1902 after signing the Boxer Protocol, which stipulated the presence of an allied military force in the capital and payment of around 450 million taels of silver in reparations to the Alliance to assure the continuity of the Qing. The consequence of the Boxer Rebellion was disastrous for the Qing state, but also pressed it to implement a new series of reforms especially modeled on the Meiji Restoration of Japan.[36] One of these reforms was to solve the shortage of drinkable water that plagued the city. Although China had suffered a humiliating defeat following the First Sino-Japanese War of 1894, this was not, contrary to the assumptions of scholars, a moment when relations between Japan and China completely soured; rather, many Chinese viewed Japan as possessing technological superiority.[37]

With the increasing influence of foreigners across a wide range of Chinese affairs during the internationalization of China after the Boxer Rebellion, the management of drinking water was intertwined with both domestic politics and imperial rivalries. During the evacuation of the Manchu Court from 1900 to 1902, the Japanese lifted the ban on well drilling and challenged the state monopoly on wells in the Inner City. Wells sunk by the Japanese met with remarkable success. For example, Japanese contractors drilled a well in the Methodist Mission Compound in the spring of 1902. This well was 210 feet deep, about 2.5 inches in diameter, lined with bamboo, and supplied a reservoir set in the ground that was 15 feet deep and 3 feet in diameter. The water was soft, and the supply drawn by hand was entirely adequate for the demands of eight American families and about five hundred Chinese.[38]

Remarkably, Japanese well drilling challenged the Manchu state's control over wells in the Inner City. In addition, a blueprint for running water in Beijing drawn by Americans brought another challenge to the Manchu's conception of water security. After the devastating Boxer Rebellion, the American Legation

35 New Policy, *Dagongbao*, May 10, 1904.(《政令新奇》，《大公报》，1904年 5月 10日，第 2版。)

36 Reynolds, Douglas. *China, 1898–1912: The Xinzheng Revolution and Japan* (Cambridge, MA: Harvard University Asia Center, 1993).

37 Paula Harrell, *Sowing the Seeds of Change: Chinese Students, Japanese Teachers, 1895–1905* (Stanford, CA: Stanford University Press, 1992).

38 Bailey Willis, *Research in China*, vol. 1, pt. 1 (Washington: Carnegie Institution of Washington, 1907), 201.

TABLE 3 Deep wells dug by the Ito Well Drilling Company in Beijing

Location	Number	Depth (feet)	Year	Relevant Data
Japanese Legation	2	231	1900	Yield over 100 liters of water per minute
Peking-Hankow Railway Yard	1		1902	
Methodist Mission Compound	1	210	1902	Bottom in black sand, bamboo casing
German Legation	1		1902	
French International Hospital	1		1903	
Belgian Legation	1	195	1903	85 liters per min, soft water
Presbyterian Mission Compound	1	200	1903	
American Board of Foreign Missions Compound	1	150	1908	
Morrison Street	1		1909	One of the first wells drilled for public use

SOURCE: GEORGE B. BARBOUR, "DEEP WELLS IN THE PEKING AREA," *BULLETIN OF THE GEOLOGICAL SOCIETY OF CHINA* 3, NO. 2(1924): 131–133.

dug two more wells during the process of reconstruction. However, the first well proved inadequate while the second, 100 yards from the first, was sunk 32 feet, at which level water poured in in such quantities that it collapsed the well and forced the drilling work to a halt.[39] Constantly frustrated by inadequate and bitter well water, the Americans pressured the Qing state by proposing the construction of a waterworks. As early as the spring of 1900, Charles Denby Jr. (1861–1938), a former American diplomat in Tianjin, proposed establishing a running water plant to facilitate drinking water and fire fighting in Beijing. Unfortunately the Boxer Rebellion suspended this proposal.[40] In June 1901,

39 Willis Bailey, *Research in China*, vol. 1, pt. 1 (Washington: Carnegie Institution of Washington, 1907), 201.

40 *Americans proposal to founding waterworks in Beijing*, April 29, 1901, no. 01-14-031-06-001, Archives at Institute of Modern History, Academia Sinica, Taipei. (中研院近史所檔案 01-14-031-06-001, 美國請在京城創設自來水案, 1901 年 4 月 29 日。)

an American entrepreneur, Gavin Millen, petitioned the Qing government to establish a waterworks to benefit the 1,000 foreigners living in the city who were vulnerable to the diseases deriving from the impure drinking water.[41] Millen assured the Manchu state that the waterworks would be constructed and run by both Manchus and foreigners, but the defeated Qing government was reluctant to cede the rights to their drinking water and turned down this proposal.[42]

To assuage the pressure imposed by Western powers and retain its control over drinking water in the Inner City, the Qing state determined to undertake political action by fully embracing and implementing Japanese well drilling as soon as the throne returned to Beijing. In August 1902, the Ministry of Commerce petitioned to advocate for and popularize well drilling, and was approved immediately by the court.[43] With the state's permission, the Ministry of Commerce established the Institute of Well Drilling to train apprentices in well boring and hired Japanese technicians as instructors and advisors. An official named Chen Bi (陳璧, 1852–1928) had contributed to suppressing the Boxer Rebellion and was an official caretaker during the evacuation of the court, thereby winning the throne's favor. In 1901, Chen Bi was promoted to magistrate of Shuntian Prefecture (順天府尹) in charge of the capital area. To supply drinking water to the newly created Shuntian Middle School, Chen Bi contracted Japanese craftsmen to drill a deep well lined with bamboo pipe whose depth reached three hundred feet.[44] Following this, in response to a call from the throne, Chen Bi established and supervised the Institute of Well Drilling in the Bureau of Craft and Technology under the Ministry of Commerce, where he served as vice-minister. He hired Japanese technicians as instructors to teach over thirty Chinese apprentices to drill wells. In addition to class lectures, they also put theory into practice and mastered the technology of well

41 *Application for 100,000 yuan to establish Beijing waterworks*, June 29, 1901, no. 03-37-002-02-003, Archives at Institute of Modern History, Academia Sinica, Taipei. (中研院近史所檔案 03-37-002-02-003, 請於北京開始自來水公司並報效銀十萬元求奏請賞收由，1901 年6 月 29 日。)

42 Even in 1907 the Spanish envoy in Beijing went directly to the Ministry of Civil Affairs to question officials and claimed to establish running water. ("Western merchants petitioned to build waterworks," *Shuntien Times*, August 8, 1907; 順天時報 1907 年 8 月 8 日第七版, 西商邀求自來水事).

43 *Qingshilu*, August 7, 1902. (清實錄 光緒28年 1902 年 壬寅 八月初七日 第 59 冊, 112).

44 "The News from Beijing," *Shenbao*, August 1, 1902. (申報 1902 年 8 月 1日 第 3 版).

drilling. After the completion of dozens of wells, these Chinese apprentices became skilled technicians who could drill wells by themselves.[45]

To promote well drilling by Chinese craftsmen, the Qing Court and political elites were mobilized to contract new wells. For example, the Third Beizi (Mongol prince) dug two deep wells in his garden (三貝子花園), and in 1907 the palace opened two foreign wells following an edict of the Empress Dowager Cixi (1835–1908) to demonstrate substantial support, which was unprecedented, for the Chinese well drillers. One well was located to the west of the Middle Sea (中海), and the other was situated in the rear of the Hall of Benevolence and Longevity (仁壽殿) in the Forbidden City. The chief engineer who drilled the well was Pan Zhaofeng, trained at the Institute of Well Drilling.[46] These two projects took three months to successfully complete and the participants were rewarded two hundred taels of silver by the Empress.[47] The Chinese crews achieved such a level of competence through their practice of well boring that it led to the drilling of additional wells in Beijing. They even managed to repair and improve a well with a decaying bamboo casing at the Belgian Legation that had been drilled by a Japanese contractor.[48]

The Institute of Well Drilling sponsored by the Manchu state succeeded in recapturing its control over well construction and began drilling state-owned wells in the Inner City. For example, six out of seven foreign wells in the community of south Dongsi were owned by the state prior to 1911 (see table 4). This led well houses without foreign wells to feel competitive pressure based on revenue. For example, water carrier Liu rented a governmental well to sell water and paid ten taels of silver monthly. However, his water business was in great deficit because of the competition with foreign wells, so he petitioned for a 50 percent reduction in his rent charge.[49]

Even though water carriers wanted to switch to foreign wells to raise profits, the underfunded Manchu state who owned the wells in the Inner City could not afford to drill many foreign wells. Although the Japanese well-drilling

45 "Memorial of Advocating and Popularizing Well Drilling in Beijing," *Shenbao*, March 4, 1905. (申報 1905 年 3 月 4 日 第 10 版 商部奏京城倡辦鑿井漸圖推廣折).

46 "Digging Foreign Wells in the Palace," *Shuntien Times*, September 29, 1907. (順天時報 1907 年 9 月 29 日 第7版 京師新聞 宮中開鑿洋井》).

47 "Reward Well Drilling Crews in Beijing," *Shenbao*, November 14, 1907. (申報 1907 年 11 月 14 日 第 10 版 獎賞藝徒鑿井北京).

48 George B. Barbour, "Deep Wells in the Peking Area," *Bulletin of the Geological Society of China* 3, no. 2 (1924): 131–132.

49 Chinese First Historical Archives, *Shenbao zhuojian dongsanzuomen nei shuijing yuezu shi*, No. 21-0301-0005, February 13, 1907. (中國第一歷史檔案館，申报酌減东三座门内水井月租事，21-0301-0005，光绪三十三年二月十三日。)

TABLE 4 Foreign wells in the south of Dongsi (東四) in the Inner City.

Well Location	Date	Ownership
Zhegusi	1902–1908	Government
Yanzifu	1902–1908	Government
Donghua gate	1902–1908	Government
East of Dongsi	1902–1908	Government
Xifeng hutong	1902–1908	Government
Shamao hutong	1909	Government
Hongxing Temple	1910	Temple

SOURCE: *THE SURVEY OF FOREIGN WELLS IN BEIJING*, J067-001-00438, APRIL 1917, BEIJING MUNICIPAL ARCHIVES. (北京市檔案館檔案 J067-001-00438 北京洋井調查，1917 年4月.)

technology used affordable outfits and easily obtainable bamboo pipe, the price of well drilling was still unfavorable. The process of constructing a single well usually took about fifty days because of Beijing's deep, unproductive water zone, while the average payment for satisfactory fulfillment of a well contract was 1,200 yuan in silver (about US$570) and could rise to a maximum of 1,900 yuan in silver.[50] For example, a well drilled in the Methodist Mission Compound in 1902 by the Japanese cost 1,200 yuan in silver.[51] Because drilling was so expensive, the bulk of foreign wells were situated at the palace, princes' mansions, new factories, legations, and missionary compounds, all of which ordinary Manchu families could not access.

More significantly, the banner families' stubborn appetite for water from traditional wells further thwarted the state's plan to turn to foreign wells. Although Beijing residents had long relied on an impure water supply from traditional wells, they were not immediately open to the "foreign water" retrieved from foreign wells. As noted above, the inhabitants, especially residents in the Inner City, relied on *erxingzi* water (half sweet, half brackish water) from traditional wells, which had gradually shaped the Manchu's taste in water and became a symbol of Manchu identity in Beijing. This preference was so ingrained that it became an obstacle to the expansion of foreign wells in the Inner City.

50 George B. Barbour, "Deep Wells in the Peking Area," *Bulletin of the Geological Society of China* 3, no. 2 (1924): 130.

51 Willis Bailey, *Research in China*, vol. 1, pt. 1 (Washington: Carnegie Institution of Washington, 1907), 201.

FIGURE 5 Pump well rig in Beijing
SOURCE: 1917–1919, SYDNEY D. GAMBLE PHOTOGRAPHS COLLECTION, DUKE UNIVERSITY LIBRARY

Facilitated by the state-sponsored Institute of Well Drilling, the state mastered and accommodated Japanese well-drilling technology in the Inner City to dilute the upsetting impact of new well drilling. In this way, the state secured its sovereignty over drinking water in the Inner City where the throne and its followers resided. When compared to the state-sponsored well drilling in the Inner City, the ways in which the water carriers in the Outer City responded and participated diverged substantially: they cooperated with merchants and patronized Japanese well drilling to seek profits. If the water regime in the Inner City was motivated by the state, the water regime in the Outer City was driven by the market, which proved to be more flexible and resilient.

4 Foreign Wells in the Outer City

While the Manchu state moved to reclaim and control the wells in the Inner City by founding the Institute of Well Drilling to promote foreign wells, the Outer City was left in the hands of the water carriers; although in 1904 the Institute of Well Drilling also drilled nine wells on government land in the Outer City as a way to master drilling technology under the instruction of Japanese technicians.[52] Water carriers began to engage in the foreign well business as early as 1901. In that year, when the Germans retreated to the Beijing Legation Quarter, a water carrier bought their well and sold the water for remarkable profits.[53] Once people had a chance to become familiar with the advantages of Japanese well drilling, local sponsors were eager to adopt the new technology. Thereafter, more water carriers, who had an acute sense of the water business, began contracting with Japanese well-drilling companies to either open new wells or deepen existing wells so that they could run their water businesses with foreign wells. Accordingly, water carriers rather than the state became important vehicles of technology transfer in the Outer City.

In the eyes of water carriers, Japanese well drilling represented technological superiority, but needed to be adapted to Beijing's environment and context. The immediate challenge came from geomancy (*fengshui*). In traditional China, the opening of a well was associated with many rituals. According to

52 Chinese First Historical Archives, *Wei guizimiao guandi xi zaojing zhiyong ying yuanzhao wucheng mianjiao zuyin qing beian shi zhi neiwufu*, No. 05-13-002-002018-0101, October 15, 1905. (中國第一歷史檔案館，为皈子庙官地系凿井之用应援照五城免交租银请备案事致内务府，05-13-002-002018-0101，光绪三十一年十月十五日。)

53 Unidentified author, *The General History of Foreign Wells in Beijing, The Trivial Record of Beijing's History*. (出版者不詳，燕京今古瑣聞錄 民國, 北京市洋井之略歷。).

the theory of geomancy, *qi*, an essential concept in *fengshui* and a movable force engendering life, would be transmitted from under the ground through improperly opened wells. An influential *fengshui* manual categorized sixteen types of wells and specified their sites, shapes, and depths.[54] Furthermore, Beijing had a well deity cult since a well is one of six necessities in the house, and a sacrifice would be offered to wells at the winter solstice (冬至). As a consequence, to adapt to Chinese practices, Japanese well drillers had to comply with these geomantic guidelines by following the advice of a *fengshui* specialist in choosing a drilling date and site. For example, in 1906, a well was sunk outside the Xizhi Gate and was covered with a piece of red cloth to ward off evil and calamity. Because the Chinese believed that a newly opened well was innocent and pure, if it became possessed by evil, it would be contaminated forever.[55] In addition to the cultural challenge of geomancy, the real challenge of Japanese well boring was to economically and socially navigate the Outer City.

Since the 1650s, Shandong water carriers in the Outer City had been sinking wells and involving themselves in water businesses, so it was natural for them to turn to foreign wells. The existing wells, customers, and networks of these well houses gave their enterprises an advantage over other potential sponsors of well digging. Although running water businesses with foreign wells was appealing, the cost of foreign well-drilling was formidable since the payment for a contract was a hundred taels of silver. Thus, water carriers usually cooperated with merchants to raise funds to contract with Japanese well drillers. In 1902, water carriers and merchants founded a joint-stock water company in the South Incense Mill (南香廠) and called it the Natural Spring Bureau. Its well had been drilled by the Japanese and produced high-quality water, as the bureau stated in its advertisement:

> This well will provide water to shops and residents outside Qianmen, who used to rely on impure water from old wells and led to many health problems. The well will start to sell water on November 8th. The water originates from an authentic underground spring and is free from surface

54 For example, the well boring had to be made by placing anxi in all four directions in the courtyard on an auspicious date. It was believed that an inappropriate well was inauspicious and would jeopardize the household; this was an ingrained notion shared by officials and residents alike in Beijing. See Li Chunfeng, *Commentaries and Illustrations of Well-Drilling*, Masters, Continuation of Imperial Collection of Four, 599–608. (李淳風: 鑿井圖經，續修四庫全書 子部 術數類 陽宅大全, 第 599–608 頁。)

55 "Hang a Red Cloth When Drilling a Well," Peking *Mandarin Daily*, no. 666, May 1906. (京話日報 六第六百六十六號 第四版 1906 年 5 月。)

> contamination, so there are no parasites at all. Furthermore, the water is assayed by a noted chemist and proves healthy, pure, and sweet.[56]

Water carriers distributed water from this well in their territories. This foreign well water carried a certification issued by the bureau that differentiated it from the adulterated water of traditional wells.

Any new water business was dependent on the existing water carriers, whose cooperation and input were essential to its success. For example, in 1903, a merchant named Zhang Dianchen in the Outer City founded the Sweet Spring Bureau, hired Japanese technicians to drill a well, and collected money in the form of shares from investors in the bureau. Its prospectus read as follows:

> The traditional wells of Beijing are unhealthy and have resulted in health problems. Now the foreign wells drilled by Japanese technology can access sweet underground spring water, which is clean and without any infiltration. Our bureau has secured official permission to drill foreign wells, and has already succeeded in opening a few wells situated at Tangzi Hutong (堂子胡同) and Tongzi Hutong (桐梓胡同) respectively. Their water is fresh and cheap, so the residents in the neighborhood have rushed to purchase it. Now the bureau plans to drill dozens of new wells to facilitate the drinking water of merchants and residents. Nevertheless, the cost, including land purchase, well house construction, well drilling, and buckets, is high, so we are determined to collect money by shares. Each share is worth two hundred taels of silver and the maximum shares held by one shareholder is fifty.[57]

This Sweet Spring Bureau could earn three to four taels of silver per day by selling foreign well water. One-quarter of its income, around one tael, was paid to water carriers. In addition, the state used taxes to control water carriers' power as one dimension of the relationship between the state and the water carriers. According to Manchu state regulations for wells in the Outer City, a well house needed to pay a rent charge of five taels of silver per well each month—even if the wells were on private property—plus a water sales tax.[58] For example,

56 "Trivial Talk from Beijing," *Shenbao*, December 4, 1902. (申報1902年 12月 4日第 2版。)

57 "The record of Beijing," *Shenbao*, November 25, 1903. (申報1903年 11月 25日第 2版。)

58 Chinese First Historical Archives, *Huomian mindi kaijing dizu*, No. 21-0301-0040, October 12, 1907. (中國第一歷史檔案館，21-0301-0040，为外城总厅申请豁免民地开井地租事给会计科移文，光绪三十三年十月十二日。)

Wang Dehai opened two wells and paid forty taels of silver from November of 1906 to February of 1907.[59]

Cost is one of most important factors determining the popularization of new technology. Because of its high cost, not all water carriers adopted Japanese well drilling enthusiastically. Up until 1905 only one water bureau with a foreign well, which was located at Langfang Hutong (廊房胡同), had turned a fair profit; the rest experienced losses and many stockholders regretted investing in foreign wells.[60] Zhang Dianchen, the owner of Sweet Spring Bureau, was jailed because of loan defaults and committed suicide while in custody.[61] As a result, according to an investigation in 1906 by the Ministry of Civil Affairs in the Right Third Zone of the Outer City (外城巡警右廳三區), none of the ten well houses owned foreign wells.[62] The Japanese well-drilling companies attempted to reduce the cost by hiring only well-sinking specialists from Japan and employing the remaining workers from Beijing. But they also acknowledged that the intimidating cost of construction impeded the expansion of business and accordingly they only targeted elites, including high officials and rich householders, as potential clients.[63] Consequently, water carriers chose to work with Japanese companies to court rich clients while supplying poorer customers with cheaper water from traditional wells.

To bring Japanese well drilling into the Outer City, the water carriers and merchants also needed to persuade the residents to accept foreign well water. Although the Chinese residents had not developed a stubborn taste for *erxingzi* water as their privileged counterparts in the Inner City had, they still harbored a deep suspicion regarding foreigners and conceived of the water procured through Japanese techniques and Chinese techniques differently, given the fresh memory of the Boxer Rebellion. What's worse, the foreign wells

---

59 Chinese First Historical Archives, *Wang Dehai zaojing an*, No. 21-0301-0034, April 1907. (中國第一歷史檔案館，民政部档案，王德海凿井案，21-0301-0034，光绪三十三年四月。)

60 "The Strategy of Selling Water," *Peking Mandarin Daily*, 1905, issue 462, no. 3. (京話日報（四）1905 年 第四百六十二號 三 賣水也得有訣竅。)

61 "Died in Dentation," *Dagongbao*, February 25, 1904, p. 2. (《羁押致死》，《大公报》，1904 年 2 月 25 日，第 2 版。)

62 Chinese First Historical Archives, "The Survey of Merchants in the Outer City of Beijing in Late Qing Period," *Historical Archives* 3, (September 2001): 70–73; and Chinese First *Historical Archives*, "The Survey of Merchants in the Outer City of Beijing in Late Qing Period," *Historical Archives* 4, (December 2001): 53–60. (中國第一歷史檔案館，清末北京外城商戶調查表（上, 中），歷史檔案，2001 年第 3–4 期, 70–73; 53–60).

63 "Japanese Well-Drilling Company," *Shuntien Times*, September 5, 1908, p. 8. (順天時報 1908 年 九月初五日 第一版 日本鑿井局。)

in the Outer City were drilled by Japanese craftsmen rather than by Chinese craftsmen as in the Inner City. Because of the common Chinese habit of using white alum, a fine powder, to purify turbid drinking water, the residents tended to believe that the Japanese also added chemicals or medicine to the water. Since the residents did not trust the Japanese, they speculated that this "foreign water," as they called it, must be destructive and detrimental to the human body.[64]

This belief was quite prevalent in China, as the following episode illustrates. He Gangde, the prefectural magistrate of Suzhou in the Yangtze Delta, had been a resident in Beijing. He tried to introduce Japanese well drilling to Suzhou, where residents relied on river water in which they also washed rice, did their laundry, and cleaned chamber pots. The magistrate managed to employ well-drilling craftsmen from Beijing to drill three foreign wells. To his dismay, the residents were reluctant to draw water from these wells. When asked, they all responded that foreign well water was poisonous, while river water was more reliable. The magistrate deplored the stubbornness of the residents.[65]

Well owners felt compelled to denounce this prevailing suspicion and reassure the Chinese residents during the popularization of Japanese well drilling by, among other things, seeking help from the newspapers. The *Peking Mandarin Daily*, a newspaper with a large circulation that targeted the lower class in Beijing, elaborated the process of Japanese well drilling and explained that foreign wells yielded clean water purely because of their depth. This allowed access to high-quality underground water that didn't require the use of chemicals or medicines.[66] The reality was more persuasive than the argument: the water drawn from foreign wells was far better than the salty water from traditional wells. People slowly recognized this and accepted foreign wells with their "sweet water." Japanese well drilling was gradually adopted in the Outer City with the sponsorship of water carriers and merchants. A foreign well survey for the South Left District of the Outer City suggests that ten out of thirteen foreign wells were private (see table 5). With the advent of running water in 1910, the bulk of water carriers were pushed into patronizing more foreign wells in order to survive this new competition from tap water, which is the subject of the next chapter.

64 *Peking Mandarin Daily* (京話日報) 12, no. 2 (1904).
65 He Gangde, *Huamengji*, vol. 1 (何剛德：活夢集 卷上).
66 *Peking Mandarin Daily* (京話日報) 12, no. 2 (1904).

TABLE 5 Foreign wells of the South Left District in the Outer City.

Well location	Date	Ownership
Dongchashi hutong	1901	Government
Dongheyan	1902	Private
Chongwen gate	1902	Private
Leijia hutong	1902	Private
Qianmen street	1903	Government
Tianqiao	1904	Government
Feichang	1904	Private
Xuandi temple	1905	Private
Zhongsitiao	1905	Private
Sanlihe	1905	Private
Hebochang	1905	Private
Dashi	1907	Private
Xilishiying	1907	Private

SOURCE: *THE SURVEY OF FOREIGN WELLS IN BEIJING*, J067-001-00438, APRIL 1917, BEIJING MUNICIPAL ARCHIVES. (北京市檔案館檔案 J067-001-00438 北京洋井調查，1917年4月。)

5 Conclusion

Following the conclusion of the Boxer Rebellion in 1900, and contrary to the assumption that western technology shaped modern China, the introduction of Japanese well-drilling technology revolutionized the sources of drinking water in Beijing and, more importantly, lifted the ban on well drilling in the Inner City. The changing political and technological circumstances threatened water security and altered water's relationship with carriers. To reassert its control, the state patronized Japanese well drilling through the state-sponsored Institute of Well Drilling and dug state wells for the water carriers in the Inner City. Water carriers in the Outer City directly participated in well drilling to perpetuate their water regime.

The spatial order created and imposed by the water regime was ever-changing under the influence of new technology and state-making. The water revolution was complicated by the ethnic schism between the banner and non-banner people that was reinforced by the state as part of the city's political and social hierarchy. In this way, the state also frustrated the unity of the water regime. The water carriers in the Outer City utilized a different mechanism for incorporating foreign technology into businesses, with economic incentives replacing the political motives of the Inner City. Moreover, contrary to the assumption made by recent studies of state-building technologies that China was a passive recipient of Japanese technological transfers,[67] this chapter challenges that view by showing that both the state and the water carriers were participants in water modernization and constantly renegotiated their roles in the daily life of Beijing. The water carriers would continue to use the foreign well water to compete with the arrival of tap water and to defend their water regime.

67 Yang Daqing, *Technology of Empire: Telecommunications and Japanese Expansion in Asia, 1883–1945* (Cambridge: Harvard University Press, 2011).

CHAPTER 3

New Water, Old Carriers, 1910–1923

The introduction of the first modern waterworks in Paris in 1802 marked a decisive shift away from Western cities' historic reliance on wells. The modernization of drinking water in Beijing was an even more complex process. The agenda of Western power in China had the unintended consequences of leading to modernization in Beijing, not least in treaty ports such as Shanghai, Hankou, and Tianjin, as illustrated by the implementation of Japanese well-drilling discussed in chapter two. Furthermore, after the Boxer Uprising, pressure from foreigners directly contributed to the completion of a new infrastructure for running water—Capital Waterworks (京師自來水公司)—in Beijing in 1910. The advent of running water and this new water infrastructure changed the dynamic of water governance in Beijing. To adapt to this challenge to their dominance, water carriers managed to rebuild and maintain their water regime by incorporating running water into it until the collaboration between Capital Waterworks and the carriers was suspended in 1923.

The other event that complicated the water regime in Beijing was the collapse of imperial rule when the 1911 Revolution overthrew the Manchu-led Qing dynasty. On 10 October 1911, an armed uprising broke out in Wuchang, which soon provoked nationwide revolts to renounce the Qing court. In February 1912, the last emperor Puyi abdicated the throne. Yuan Shikai (袁世凱, 1859–1916), the leader of the Beiyang Army, became President of the nascent Republic of China and established Beijing as the capital. The 1911 Revolution eliminated the dichotomy between the Inner City and Outer City that had been upheld by the Manchu state since 1644 and it reshaped the city's spatial pattern by ending the inequality in drinking water between the banner and non-banner people. Instead of a city that could pass as the ancient capital of an old empire, Beijing was adopting new ways. Electric lights and running water, paved streets and sewer systems, railway trains, automobiles, and even airplanes could be found in the city. More and more, Beijing was the capital city of a modern republic.[1] In the midst of the struggle between the old and the new, the water carriers as hand laborers were faced with the problem of accommodating the impersonal methods of production in industry. More importantly, the water

1 Sidney D. Gamble, *Peking: A Social Survey* (New York: George H. Doran Company, 1921), 50.

 | DOI:10.1163/9789004720800_005

regime needed to adapt to the rising consciousness of public goods that was promoted by the new republican state.

1 The Foreign Proposal for Running Water

In the aftermath of the Second Opium War, the Treaty of Tianjin in 1859 granted Britain, France, the United States, and Russia residence in Beijing. This led to the building of the Legation Quarter, a section of the city set aside for the representatives of Western powers to the Imperial Court of China. The Western ambassadors brought their families and lived and worked there. The Legation Quarter later came to rival the Forbidden City as the focal point of power in the Chinese capital and even in the state. However, at the end of the nineteenth century the infrastructure of Beijing was in a state of miserable decay and was becoming unacceptable in the new world.[2]

One insufferable drawback to life for Westerners in Beijing was the foul water from the wells. Owing to the fact that the available well water was limited and inferior, the foreign inhabitants who were obliged to drink it would first place the water in large jars so that the impurities could settle to the bottom and be filtered out. Carbonation was also one of the methods employed by Westerners in Beijing to purify the city's drinking water.[3] In addition, to obtain extra pure water, foreign families in Beijing even dug large cisterns into which the water from their roofs drained, and from which a plentiful supply of soft water was available throughout the year.[4] Until 1910, when Capital Waterworks opened, Westerners in Beijing relied on these "primitive" means of obtaining water from local wells and supplemented this water with rain.

Thus, the initial impetus to transform Beijing's water supply arose out of pressure from Westerners living in the legations. Their experience in their native countries not only led them to expect access to running water but to cultivate an interest in the pursuit of profit in Beijing's water supply, especially

2 Elisabeth von Heyking, *Memoirs of Four Continents, 1886–1900,* cited in Felicitas Titus, *Old Beijing: Postcards from the Imperial City* (Tokyo: Tuttle Publishing, 2012), 49.

3 Seltzogene or gasogene is a late Victorian apparatus for producing carbonated water, using a mixture of tartaric acid and sodium bicarbonate that reacts to produce carbon dioxide to create sparkling water. Since these two chemical substances were unavailable in China, this device was not popular in Beijing. Moreover, the carbonation was unable to remarkably improve the water quality. See Mary Hooker, *Behind the Scenes in Peking, Being Experiences during the Siege of the Legations* (London: John Murray, 1911), 146.

4 John Dudgeon, *Diet, Dress and Dwellings of Chinese in Relation to Health* (London: William Clowes and Sons, Limited, 1884), 296–97.

after signing the Boxer Protocol of 1901. These foreigners, with their distinctly more advanced material civilization, viewed the upgrading of the water supply as both a necessity and a business opportunity. In the spring of 1900, Charles Denby Jr., a businessman and former American diplomat, proposed to the Zongli Yamen (Foreign Office 總理衙門)[5] that a waterworks be built in Beijing to provide water for drinking and firefighting. The Zongli Yamen put this on its agenda, but the outbreak of the Boxer Rebellion put a halt to the proposal. In April 1901, Denby resubmitted his petition to the newly established Foreign Affairs Ministry (外務部), but the Minister, Prince Qing (Yikuang, 1838–1917), postponed the proposal on the pretext of its complexity.[6]

In June 1901, another America businessman, Gavin Millen, proposed to the Foreign Affairs Ministry the construction of a waterworks. The main argument he provided in his petition letter was that the one thousand foreign residents in the city suffered from sickness and epidemics because of the lack of pure and fresh drinking water. For this reason, all foreigners in the city required tap water.[7] In anticipation of anti-foreign sentiment, the Americans repeatedly assured the Ministry that running water would be managed jointly by Chinese and foreigners. However, this was the classical period of imperialism against weak states; and thus, the defeated Qing government was anxious about ceding its monopoly over drinking water to Western powers and placing their lives in the hands of foreign countries. It is therefore unsurprising that both proposals were rejected. However, the Westerners continued to press this issue and called upon the state to build a modern waterworks capable of delivering clean and constant water.

In 1904 one foreign essayist in *The North China Herald* criticized the Qing state's persistence in monopolizing the drinking water supply while forcing the poorer residents of the capital to content themselves with nitrate-impregnated surface water.[8] In 1907 the Spanish envoy specifically went to the Ministry of Civil Affairs to question two vice-ministers about the construction of a waterworks. The angry envoy claimed that his countrymen would build a running water

5 The Zongli Yamen, established by the Qing state in 1861, was in charge of foreign policy and affairs, and was replaced with the newly created Foreign Affairs Ministry in 1901 because of its failure to deal with the Boxer Rebellion.

6 *Americans proposal to found waterworks in Beijing,* April 29, 1901, no. 01-14-031-06-001, Archives at Institute of Modern History, Academia Sinica, Taipei. (中研院近史所檔案，01-14-031-06-001，美國請在京城創設自來水案，1901 年 4 月 29 日。)

7 *Application for $ 100,000 to establish Beijing waterworks,* June 29, 1901, no. 03-37-002-02-003, Archives at Institute of Modern History, Academia Sinica, Taipei. (中研院近史所檔案，03-37-002-02-003，請於北京開始自來水公司並報效銀十萬元求奏請賞收由，1901 年 6 月 29 日。)

8 "The Making of a New China," *The North China Herald,* October 30, 1903, 894.

system for the city even if the court opposed it.[9] Ultimately, the Westerners' request for running water compelled the Qing court to plan the construction of a waterworks in order to solve the problems of insufficient and impure water service, and more importantly, to retain its control of drinking water.[10]

2 The Construction of Capital Waterworks

A waterworks affects not only drinking water but other water needs in urban areas, including water for the purposes of sanitation, firefighting, industry, and original public utilities. The Qing state initially left water procurement in Beijing for various uses to individuals who secured it from wells, ponds, or moats before the rise of running water. In order to meet modern expectations and address daily needs, centralized waterworks had already been introduced in the treaty ports in China during the late nineteenth and early twentieth centuries. For example, the Shanghai Waterworks Co. Ltd. was established in London in 1880 to build a waterworks in the Shanghai International Settlement. Supervised by the chief engineer John William Hart, Shanghai opened the first waterworks in China in 1883 with British funds,[11] followed by Tianjin in 1898, Guangzhou (Canton) in 1905, and Hankou in 1907. Under the influence of these treaty port cities, the capital gradually came to realize the importance and necessity of running water.

In 1904 an influential local newspaper, the *Peking Mandarin Daily*, reported that a Chinese businessman planned to establish running water in Beijing by using the noted Jade Spring as the source, because the drinking water distributed from wells by water carriers was unhealthy and expensive.[12] It is not surprising that this attempt was in vain since Jade Spring water was private imperial water and inaccessible to the public during the Qing period. In 1905 a prefectural magistrate proposed the establishment of a waterworks in Beijing to the Ministry of Commerce, but the powerful Grand Councilor Prince Qing opposed it on the grounds that the political and social atmosphere of Beijing was conservative and modern running water would disturb local residents; he thus thwarted this plan.[13] Another Chinese merchant, surnamed Ma, also petitioned to build a waterworks, but the capital he relied on was loans

9 *Shuntien Times*, August 8, 1907. (順天時報 1907 年 8 月 8 日第 7 版, 西商邀求自來水事。)
10 *Shuntien Times*, July 28, 1907. (順天時報 1907 年 7 月 28 日 第 7 版 擬自辦自來水。)
11 John W. Hart, "Shanghai Water-Works," *Minutes of the Proceedings*, ICE, 100, January 1890, 217–245.
12 *Peking Mandarin Daily* 83, no. 2 (1904). (京話日報 1904 第 83 號二 籌辦自來水。)
13 *Peking Mandarin Daily* 130, no. 2 (1904). (京話日報 1904 第 130 號二 北京自來水難辦。)

from British and German companies. Accordingly, the petition was rejected immediately.[14]

Eventually, the construction of a running water system was the result of external pressure, but it also arose from internal political demands during the reforms that followed the humiliating defeat of the Boxer Rebellion in 1900. Under the Empress Dowager Cixi, the Qing court instituted reforms in politics, the economy, the military, and culture in order to keep the Manchus in power. These reforms laid the groundwork for the modern infrastructure of Beijing.[15] The final impetus for establishing a waterworks system was the frequent fires that broke out in the crowded capital city.[16] The absence of available water in the streets led to many casualties and huge destruction of property in the conflagrations. The anxious Empress Dowager Cixi consulted Grand Councilor Yuan Shikai, the most powerful military and political figure in the last days of the Qing dynasty, about effective measures for firefighting. The solution Yuan offered was a regular supply of piped water. As a result, Yuan was appointed to oversee the implementation of running water in Beijing.[17]

In March 1908, the Qing throne finally approved a proposal for running water and initiated the project of constructing Capital Waterworks under a petition by Yang Shiqi (楊士琦, 1862–1918), the Vice Minister of the Ministry of Agriculture, Industry, and Commerce and a close associate of Yuan Shikai.[18] Yuan recommended Zhou Xuexi (周學熙, 1866–1947) to supervise the project. Zhou was a competent official and an influential entrepreneur in north China who had sponsored the well-known Kaiping Mines and the Chee Hsin Cement Company under Yuan Shikai's patronage.[19] Zhou Xuexi headed and organized Capital Waterworks as an official-merchant coalition using the model of government supervision and merchant management, a managerial approach

14 *Shuntien Times*, March 19, 1908. (順天時報 1908 年 3月 19 日 第七版 駁辦京師自來水。)

15 Douglas Reynolds, *China, 1898–1912: The Xinzheng Revolution and Japan* (Cambridge, MA: Harvard University Press, 1993).

16 Regarding the historiography of cities on fire, see Mark Tebeau, *Eating Smoke: Fire in Urban America, 1800–1950* (Baltimore: Johns Hopkins University Press, 2012); Greg Bankoff, Uwe Lübken, and Jordan Sand, eds., *Flammable Cities: Urban Conflagration and the Making of the Modern World* (Madison: University of Wisconsin Press, 2012).

17 Zhou Shuzhen, *Zhouzhian xiansheng biezhuan*, 31. (周淑貞，周止庵先生別傳 第 31 頁 京師自來水公司); see also, Stephen R. MacKinnon, *Power and Politics in Late Imperial China: Yuan Shi-kai in Beijing and Tianjin, 1901–1908* (Berkeley: University of California Press, 1980).

18 *Qingshilu*, vol. 59, 783. (清實錄 光緒 34 年 1908 年 3 月 18 日 59 冊 783 頁。)

19 Albert Feuerwerker, *Industrial Enterprise in Twentieth-Century China* (Berkeley: University of California Press, 1967), 5–28.

employed in the Self-Strengthening Movement (1861–1895).[20] This model was a form of business that recruited private capital to create enterprises. To denounce Confucian prejudice against mercantilism and to nudge the state to stimulate private enterprise, the government offered licenses and provided funds or loans to merchants building enterprises under this model. When necessary, the government also adopted measures to suppress competition and help these enterprises monopolize business. Meanwhile, the government asserted its control over these enterprises while underwriting the entrepreneurial risks.[21]

This model of government supervision and merchant management was consistent with the state's method of water governance in Beijing since 1600s, which was to place the water supply in the hands of a private agency. In the case of Beijing's running water, the reason for governmental supervision was the high cost, which included the waterworks itself as well as water pipes and water faucets. Furthermore, as a new phenomenon, running water needed government supervision and advocacy in the politically and socially traditional context of Beijing.[22] With the throne's approval for selling shares to raise funds, Capital Waterworks quickly collected three million Chinese yuan. Because of the throne's deep distrust of foreigners, the waterworks denied shares to foreigners, even though, ironically, the waterworks completely relied on imported technology, equipment, and experts from the Arnhold & Karber & Co., a Germany-registered company with a branch in Tianjin and a contracted waterworks there.

Under the guidance of Zhou Xuexi, the construction of the waterworks began in May 1908 and secured its water supply from the Sun River (孫河) located ten miles east of the city. Using river water as a running water source was not uncommon in the early twentieth century. For example, in 1913, 60 percent of London's running water supply came from the Thames River, while 20 percent was from the Lee River.[23] A water plant that included water purification with three settling ponds, ten sand-filtering ponds, and a pumping station

20 The Self-Strengthening Movement was a reform to modernize China's heavy industry and military in order to preserve Qing rule in the late nineteenth century.

21 Wellington K. K. Chan, *Merchants, Mandarins, and Modern Enterprise in Late Ch'ing* (Cambridge, MA: Harvard University Asia Center, 1977), 235–43.

22 "Nonggongshangbu wei chouni chuangshe jingshi zilaishui gongsi dagai banfazhe," (April 1908), in Beijingshi danganguan, ed., *Beijing zilaishui gongsi dangan shiliao* (Beijing: Beijing yanshan chubanshe, 1986), 1–2. (農工商部奏爲籌擬創設京師自來水公司大概辦法折（1908 年 4 月 28 日）北京市檔案館等編，北京自來水公司檔案史料 北京：北京燕山出版社，1986 年，第 1–2 頁。)

23 John Hassan, *A History of Water in Modern England and Wales* (Manchester: Manchester University Press, 1998), 22.

was set up on the bank of the Sun River to extract and treat water. The purified water was then circulated to the city through pipelines using a steam engine.

To convey water from the Sun River to the city, Capital Waterworks built ten miles of brick water mains. The construction of these long-distance water aqueducts involved overcoming unanticipated obstacles. The first impediment was from an imperial clan whose cemetery happened to be located on the pipe route. The imperial clan believed the layout of the pipeline would jeopardize the *fengshui* (geomancy) of its cemetery and therefore vehemently opposed the plan. To avoid resistance, the Waterworks shifted its route one-half mile to the south in order to detour around the cemetery. Nevertheless, the imperial clan compelled the pipeline project to be suspended for two months before a solution was achieved with the intervention of the Imperial Clan Court (宗人府).[24] The second obstacle was the funeral and mourning for the Empress Dowager Cixi and the Guangxu Emperor. In November 1908, the Emperor died one day before the Empress Dowager. These imperial funerals were important political and ritual events observed by the whole empire. To facilitate the funeral procession, the Waterworks was ordered to refill and level all the trenches dug in the streets for pipe laying. As a result, the waterworks project came to a halt and did not resume until the end of February 1909.[25]

In June 1909 the construction of the waterworks and the installation of pipelines were finally completed.[26] As an integral part of the running water system, the Waterworks built a branch plant with several pure water reservoirs outside the Dongzhi Gate northeast of the city to supply running water. At the plant, German craftsmen built a water tower at an elevation of 165 feet to exert water pressure for this gravitation system. This tower reservoir dominated the skyline of the city; from its top, one could even see over the Forbidden City, so it was not open to the public.[27] The water tower, from where water flowed to its distribution points through a complex system of pipes, became a new

24 "Zhou Xuexi jiu suowei gongsi anshe shuiguan yingpao feimu piyao gaoshi," (March 1, 1909), in Beijingshi danganguan, ed., *Beijing zilaishui gongsi dangan shiliao* (Beijing: Beijing yanshan chubanshe, 1986), 22. (周學熙就所謂 "公司安設水管硬刨墳墓" 辟謠告示（1909 年 3 月 1 日），北京市檔案館等編，北京自來水公司檔案史料 北京：北京燕山出版社，1986 年，第 22 頁。)

25 "Neicheng xunjing zongting wei anzang guangxu zanting anguanshi yu zilaishui gongi wanglaihan," (March 1909), in Beijingshi danganguan, ed., *Beijing zilaishui gongsi dangan shiliao* (Beijing: Beijing yanshan chubanshe, 1986), 23–24. (內城巡警總廳爲安葬光緒暫停安管事與自來水公司往來函（1909 年 3 月），北京市檔案館等編，北京自來水公司檔案史料 北京：北京燕山出版社，1986 年，第 23–24 頁。)

26 *Qingshilu*, vol. 60, 308–309. (宣統元年 1909 年 已酉 六月二十一日 清實錄 60 冊 第 308–309 頁。)

27 Zhou Shuzhen, *Zhouzhian xiansheng biezhuan*, 31. (周淑貞，周止庵先生別傳，第 31 頁 京師自來水公司。)

FIGURE 6 Water tower outside Dongzhi Gate with the city wall in the background
SOURCE: BEIJINGSHI DANGANGUAN, *BEIJING ZILAISHUI GONGSI DANGAN SHILIAO* (BEIJING: BEIJING YANSHAN CHUBANSHE, 1986.)

landmark and also a symbol of Beijing's conspicuous modernity. In February 1910, Capital Waterworks, drawing on the waters of the Sun River, began to supply running water to the city.

Given the extremely high price of pipes and faucets, extending water service throughout the city called for an additional outlay of capital to lay pipes and increase steam pumping. Lacking modern steel technology and steelworks, China was unable to manufacture its own water pipes and faucets in the early twentieth century; instead, these were imported from Germany, so the Waterworks was unable to bring water to individual households. The best compromise was the installation of public hydrants with faucets in the streets, rather than internal water fittings in each household. This arrangement was very similar to the water supply in American cities between 1840 and 1870. In both cases, urban dwellers obtained water from publicly or privately owned, centrally located waterworks that distributed water to residents through open hydrants and public faucets. But in the end, only their scale and organization distinguished these centralized waterworks from the public wells and cisterns they had replaced.[28] In February 1910, Capital Waterworks founded ten branch bureaus in different parts of the city and installed household faucets for the wealthy. To supply a great many other families, the Waterworks installed 420 public standpipes in the main streets and assigned its men to sell water. Each standpipe had a clearly visible sign with hours of business, the water price, and instructions for purchasing water.[29] Residents who wanted to use running water had to purchase tickets issued by the Waterworks at the branch bureaus in exchange for water from these street hydrants, and then they had to carry the water home themselves.

To sell its product, the Waterworks utilized two concepts—sanitation and nationalism—to advertise and popularize running water. The hygienic notion of drinking water in Beijing was heavily influenced by medical missionaries, who were the first to use modern science and technology to assay and evaluate drinking water in the city. Throughout his long residence in Beijing, Dr. John Dudgeon (1837–1901) paid attention to drinking water by studying the residents' health-related habits and composing semiannual reports for the Chinese Maritime Customs Service.[30] The institutionalization of sanitation in Beijing was rooted in the reforms in the first decade of the twentieth century

28 Maureen Ogle, *All the Modern Conveniences: American Household Plumbing, 1840–1890* (Baltimore: Johns Hopkins University Press, 1996), 36.

29 *Shuntien Times,* March 6, 1910. (順天時報 1910 年 3 月 6 日 第七版 賣水標牌。) Also, see John B. Grant, *Report on a General Health Survey of Peking, China,* February 1922, RF RG 5 Series 2, SS-601, Box 55, F346, Rockefeller Archive Center, New York, 30–31.

30 Dr. Dudgeon was a Scottish physician who received his MD from the University of Glasgow in 1862. The next year he was appointed to the Medical Mission of the London Missionary Society (LMS) to serve at the hospital in Beijing. "Obituary: John Dudgeon, M.D.," *The British Medical Journals* 1, no. 2098 (March 16, 1901): 679.

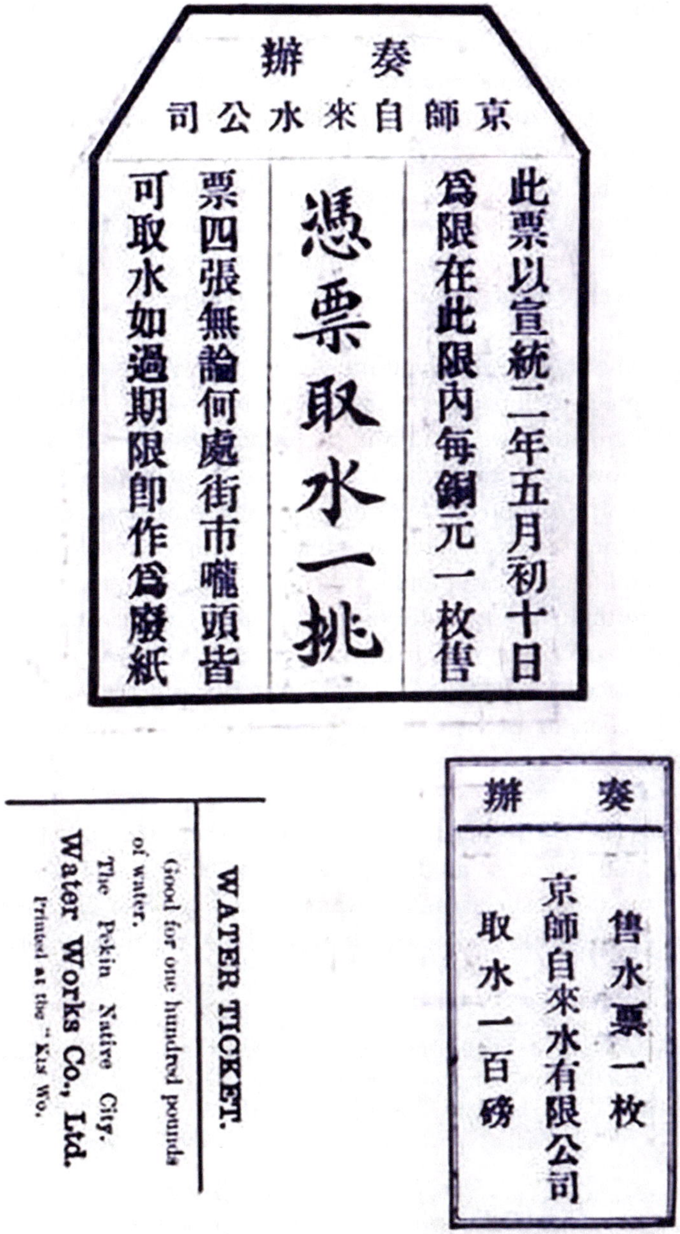

FIGURE 7 Running water tickets (1910)

under the huge influence of the Japanese public health system.[31] In 1906 Inspector Tan of the Sanitation Bureau suggested publishing a newspaper to propagate public health.[32] In 1908 a newspaper titled *Hygiene Mandarin News* was published to popularize sanitation. One of its articles elaborated the problem of drinking water that came from wells and rivers, and suggested that the authorities supervise drinking water.[33] In the new penal code of the Qing state, Article 24 declared that anyone supplying filthy and contaminated drinking water was guilty of a felony punishable by fifth-class imprisonment or a fine of up to 100 *yuan*.[34] Concerns regarding sanitation were a factor contributing to the establishment of a waterworks in Beijing in 1909.

The Waterworks highlighted the sanitation of running water by articulating the scientific treatment it had undergone. The Waterworks pointed out that the first precaution taken was to ensure a pure supply by building dams to inhibit the growth of vegetation in the upper Sun River. In addition, the Waterworks strove to restrict human activities to protect the watershed from contamination. The second precaution was the scientific treatment of running water through the use of slow filters that incorporated a deep filtering medium of up to several feet of gravel and sand through which water passed before distribution to the city.[35] In order to further prove the quality of its sanitation, the Waterworks publicized in the newspaper the assay report for its running water. It also appealed to the Capital Police Department, which agreed to advise city dwellers to use tap water rather than dirty well water. The proclamation stated:

> Drinking water is a daily necessity; if it is inferior or impure it can trigger disease and plague, so all the major cities in eastern and western countries have established running water companies that choose the best source of water, filter and purify it by machine, and then supply it to the

31 Wataru Iijima, "Jindai Beijing gonggong weisheng zhi kaocha-20 shijie chuqi de zhongguo guojia weisheng," in Beijingshi danganguan bian, *Dangan yu beijingshi guoji xueshu taolunhui lunwen* (Beijing: Zhongguo dangan chubanshe, 2003), 68–74. (飯島涉：《近代北京公共衛生之考察—20世紀初期的中國"國家衛生"》,北京市檔案館編：《檔案與北京史國際學術討論會論文》(下冊)北京：中國檔案出版社，2003年，第68–74 页。)

32 *Jinghua ribao*, April 1906. (京話日報 1906 年 第四版 講求衛生。)

33 *Weisheng baihuabao*, July 1908. (衛生白話報 第1年第 3 期 1908 年 7 月，家庭的衛生。)

34 "Daqing xinxinglu," Zhang Daqing, *Zhongguo jindai jibing shehuishi, 1912–1937* (Jinan: Shandong jiaoyu chubanshe, 2006), 86. (《大清新刑律》1907, 張大慶：《中國近代疾病社會史，1912–1937》濟南：山東教育出版社，2006 年，第 86 頁。)

35 *Shuntien Times*, February 2, 1910. (順天時報 1910 年 2 月 2 日 第七版 京師新聞 自來水之化驗。)

> residents through iron pipe. The water they provide is sweet and pure and convenient to consume, and it can be used for firefighting.
>
> In contrast, the drinking water in Beijing is procured from wells, and the majority is brackish because of impurities. Moreover, the construction of wells is unscientific. Many of them are dilapidated and go without annual cleaning, and as a result, rain and dust easily flow into the wells and do great harm to the sanitation. Accordingly, we recommend using tap water instead of well water to avoid jeopardizing your health.[36]

The second concept used by the Waterworks to advertise and popularize running water was the concept of "national water." The Waterworks claimed that its product was national water instead of foreign water. In the first decade of the twentieth century, in the aftermath of the Boxer Rebellion, the state and society in Beijing were still suspicious of foreigners. The shrewd water merchants mobilized this sentiment to their advantage. The Waterworks declared that it operated under imperial consent and, more importantly, relied entirely on Chinese capital and employed Chinese staff. In other words, the new waterworks boasted that it was producing "national water," which was completely China-made.[37] The notion of national water was vividly exemplified in the Waterworks' newspaper advertisements. One advertisement in 1910 stated: "Our Company operates this running water business with the consent of the throne, relying one hundred percent on Chinese capital and Chinese staff. The Company is not only seeking profits, but is also concerned with the public good. All enlightened dwellers favor our national water, while only ignoramuses would want to call it 'foreign water.'"[38]

Employing the two notions of sanitation and national water, the Waterworks legitimized its product and expected the city dwellers who had long abided impure supplies to greet the arrival of running water with enthusiasm; but it underestimated the extent and intensity of the opposition the running water aroused, and never foresaw the battle that lay ahead. Moreover, because the establishment of Capital Waterworks grew out of a combination of foreign

36 *Shuntien Times*, March 6, 1911. (順天時報 1911 年 3 月 6 日 第七版 內外城巡警總廳爲出示勸諭事。)

37 For the reference to "national water" that "was completely China-made," see Karl Gerth, *China Made: Consumer Culture and Creation of the Nation* (Cambridge, MA: Harvard University Asia Center, 2003).

38 "Jingshi zilaishui gongsi gaobai," (January 1910), in Beijingshi danganguan, ed., *Beijing zilaishui gongsi dangan shiliao* (Beijing: Beijing yanshan chubanshe, 1986), 61. (京師自來水公司告白 1910 年1月，北京市檔案館等編, 北京自來水公司檔案史料 北京：北京燕山出版社，1986 年，第 61 頁。)

pressure and Chinese political strategy, rather than from consistent internal water demands based on growing industrial and domestic needs, it was thus vulnerable to political fluctuation and faced an unpredictable future.[39]

Prior to the mid-nineteenth century, in major world cities equipped with running water, the delivery of water was considered a private good and was organized by firms operating in a competitive market environment. From the late nineteenth century onward, cities began to municipalize their water supplies. Taking England between 1861 and 1881 as an example, a number of factors account for this trend. First, water provision enjoyed economies of scale and its social value exceeded its private value. In combination with the natural monopoly characteristics of the water supply, the externality problem tended to favor public control over the provision of water. Second, the municipal water enterprise appeared to be responding to community needs better than most private companies had ever done.[40] This pattern of water municipalization never held true in China. The Capital Waterworks was, as claimed by its shareholders, constructed as a private enterprise: "The Peking Water Works was entirely financed by the commercial Community."[41] This claim also suggested that the municipal government never intended to take responsibility for municipalizing drinking water as a public good. This led to a dilemma in the following decades, because merchants used the Waterworks to seek private profits while the government tried to favor public goods.

The predicament of the state's water governance was illustrated by Zhang Binglin (章炳麟 1868–1936), a renowned scholar and revolutionary in the early twentieth century. In 1914 Zhang was put under house arrest in the Longquan Temple outside Xuanwu Gate (宣武門) for criticizing President Yuan Shikai. Unfortunately, as a native of southern China, Zhang was not used to the temple's unsavory well water and had diarrhea, a symptom with a long history in China's well water saga. The Ministry of Internal Affairs instructed the Waterworks to install running water in the temple for Zhang and also for two hundred orphans nursed by the temple, but the Waterworks refused this order on the grounds that the temple was too distant from the water mains and the cost

39 In the case of New York City, the 1830s saw increasingly frantic lobbying by chemical works, breweries, tanneries, distilleries, hotels, sugar refineries, and other parts of the fast-growing New York economy who relied on pure and reliable water supplies and who would be ruined if the city authorities failed to take action. See Matthew Gandy, *Concrete and Clay: Reworking Nature in New York City* (Cambridge, MA: MIT Press, 2002), 5.

40 R. Millard, "Privatisation in Historical Perspective: The UK Water Industry," in D. Cobham, ed., *Money, Trade and Payments: Essays in Honour of Dr. Coppock*, 197 (Manchester: Manchester University Press, 1989).

41 "The Peking Water Works," *Peking Daily News*, April 30, 1914, p. 4.

of installing pipe would reach as high as 1,000 *yuan*.[42] The misfortune of Zhang Binglin exemplifies the Waterworks' prioritization of its private interests over the public good and the limited influence of government over the Waterworks.

3 The Running Water Regime

The modernization of the water industry brought better piped supplies and reshaped the spatial pattern of Beijing. The first aspect of the spatial remaking of running water was the division between the eastern city and the western city. With the demise of the Qing dynasty in 1911, reform and revolution had virtually erased the dichotomy between the Inner City and Outer City that had been sponsored by the Manchu state since 1644. Nevertheless, the uneven distribution of running water created a new spatial inequality in drinking water: running water was more popular in the eastern section of the city, which was closer to the water tower, but less accepted in the western section of the city because of the lower water pressure there. As figure 8 indicates, the diameter of mains in the east was larger than their counterparts in the west. As a result, the spatial dichotomy in drinking water switched from a pattern of Inner-Outer City in the Qing dynasty to a pattern of Western-Eastern City in the Republic, and the city's social inequalities shifted to a western-eastern difference as well.

The second aspect of reshaping the spatial pattern of Beijing was that the distribution of running water divided the city into piped areas and unpiped areas. One factor that accounted for this spatial inequality was the extremely high price of pipes and faucets. To make ends meet, the Waterworks needed to maintain at least twenty thousand households with water fittings installed. However, the municipal government could not contribute to installation costs and could only exhort residents to install the pipe at their own expense.[43] Despite the fact that the Waterworks had attempted to expand its pipe network and make running water accessible to more private households, the installation cost of branch pipes and faucets was so high that even for households in the same district as the main pipe, only public buildings and wealthy households could afford the installation. In 1910 the Shuntian municipal office installed running water as a symbolic gesture to advocate its use, but it cost more than

42 "Longquansi zilaishui an" [The case of running water of Longquan Temple], 22 March 1914, J181-18-21737, Beijing Municipal Archives.

43 Zhongguo diyi lishi danganguan, "Qingmo Beijing kuochong zilaishui zhuanguan shiliao," *Lishi dangan*, no. 2 (1992): 64. (中國第一歷史檔案館，清末北京擴充自來水專管史料，歷史檔案 1992年第 2 期，第 64 頁。)

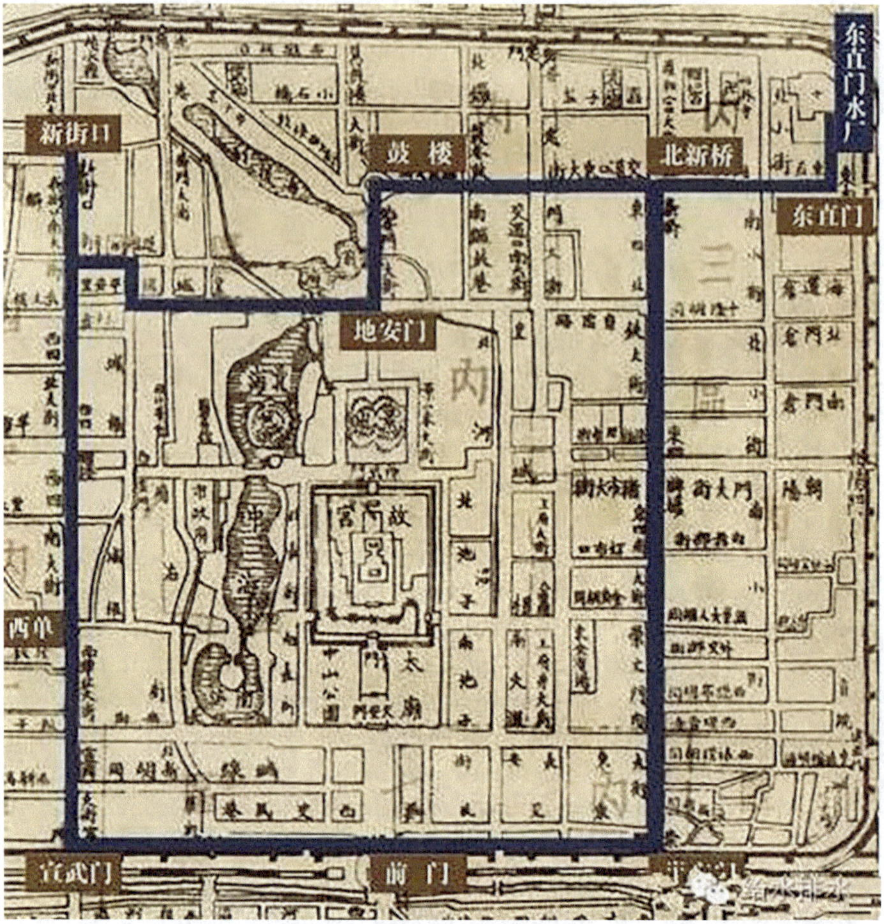

FIGURE 8 A map of the running water mains in the Inner City in the 1910s, based on *Beijing zilaishui gongsi dangan shiliao* (Beijing: Beijing yanshan chubanshe, 1986.)

one hundred taels of silver.[44] Spatial inequality is a proxy for social inequality. Access to running water service was unevenly distributed throughout the city because the network for the waterworks was concentrated in the commercial and affluent residential districts of the city, while the more remote and poorer communities were provided with less water or were even left unserved.[45] Since

44 *Shuntien Times*, March 14, 1910. (順天時報 1910 年 3 月 14 日 第七版 京師新聞 順天府安設自來水。)

45 This was common for nineteenth-century American cities. Historian Robin L. Einhorn calls it the "segmented system," a system that provided benefits to those who paid for them but which also made the American urban landscape a physical expression of

water never extended to any but the wealthiest homes, most residents were unable to experience the true convenience of water in the household.

Coupled with the expensive installation price, another factor that contributed to the formation of unpiped areas was the high rate of temporary migration and mobility in Beijing's neighborhoods. Especially in the Outer City, migrants from the empire, including bureaucrats, merchants, and students, constituted a high percentage of the total population. As a result, a large number of local residents made their living by subleasing their houses (吃瓦片), but landlords were disinclined to assume responsibility for installing running water for tenants since doing so would require raising rents. The tenants, in turn, were reluctant to invest any money to improve the facilities, albeit some renters would install internal faucets. For example, the Xiao family lived in a rental house in the western part of the city and installed running water pipes at their own expense. When they moved out, they asked Capital Waterworks to relocate these expensive items to their new home but were stopped by the former landlord, which led to a disputation.[46]

In the piped communities, the arrival of running water imposed an acute threat to the water carriers' business by intruding into their distribution territories, which were an exclusive space created by the water carriers to dispense water to households. The public and private faucets installed by the Waterworks presented direct competition to the water carriers and aroused their vehement resistance. Water carriers immediately responded to these challenges by defending their water regime. The simple weapon they employed was the sabotage of public faucets, which were connected to hydrants by only one iron thread and were therefore susceptible to theft and vandalism.[47] Although the Waterworks assigned keepers to the street hydrants, their duties were limited to daytime hours. These unattended public hydrants provided the resentful water carriers with a great opportunity for retaliation at night. The vandalism of faucets caused huge losses for the Waterworks since all the faucets were imported from Germany and were expensive to replace; moreover, vandalism severely jeopardized the selling of running water. As a consequence, the helpless Waterworks had to lock its hydrants and petition the authorities to secure police protection.[48]

---

political inequality; see Robin L. Einhorn, *Property Rules: Political Economy in Chicago, 1833–1872* (Chicago: University of Chicago Press, 1991), 104.

46 *Shuntien Times*, May 4, 1917. (順天時報 1917 年 5 月 4 日 第七版 自來水管糾葛。)

47 *Shuntien Times*, February 12, 1909. (順天時報 1909 年 2 月 12 日 第七版。)

48 *Qingshilu*, vol. 60, 308–309. (宣統元年 1909 年 已酉六月二十一日 清實錄 60 冊 308–309 頁。) The theft of faucets was a chronic problem in Beijing. For example, the faucets in three streets outside Xuanwu Gate were stolen in one night in 1918. See *Chenzhong bao*, May 22, 1918. (晨鐘報 1918 年 5 月 22 日 第六版 水管鐵器被竊。)

In parallel with this sabotage, the other strategy utilized by water carriers was to exploit the notion of foreign water to catalyze nationalism-based resistance. The Waterworks deftly manipulated the sentiments of nationalism in its business to create the idea of "national water," and it legitimized this idea through the consent of the throne and exclusive reliance on Chinese capital. However, there was, in fact, solid evidence to show that running water was foreign water. First, the emperor himself relied on Jade Spring water rather than on running water. The Forbidden City did not install running water until 1914, three years after the Qing emperor abdicated.[49] Second, the water enterprise was entirely contracted out to Germans for equipment, technology, and craftsmen. Both of these gave wary water carriers and residents an excuse to question the idea of national water, which, in reality, seemed all too foreign to be good, especially for the banner people in the Inner City who were still habituated to *erxingzi* water.

Water carriers furthered this mentality by propagating the idea that running water, as foreign water, had been dosed with medicine, because running water was full of tiny bubbles when it streamed out of a hydrant into a bucket. Moreover, the water carriers claimed that running water came from underground pipelines without exposure to sunshine, which meant that it was Yin-water (negative water) according to the theory of *fengshui*. Drinking this kind of water was thought to cause a person to accumulate negative vital energy and, ultimately, to endanger the human body. The Waterworks had to run advertisements in newspapers to explain the bubbles in the tap water.[50] The Waterworks also held an exhibition under the water tower that aimed to demonstrate that the production methods for running water were scientific and the water quality was high. Such efforts were needed in order to prove that the allegations about running water using medical powder in its purification, or that the pumping machine was fueled by children's blood, were groundless and ridiculous.[51] Additionally, to appeal to residents, the Waterworks promoted the water at half price for three months.[52]

In spite of these measures, running water had two inherent problems in its daily operations. The first problem stemmed from a lack of hydrant keepers, who had to endure all kinds of weather without shelter, from the bitter cold

49 *Aiguo baihuabao*, June 4, 1914. (愛國白話報 1914 年 6 月 4 號 第四版 內廷安設水管。)

50 *Shuntien Times*, February 3, 1910. (順天時報 1910 年 2 月 3 日第七版 定期賣水。)

51 *Shuntien Times*, March 19, 1910. (順天時報 1910 年 3 月 19 日 第七版。)

52 "Jingshi zilaishui gongsi gaobai," (January 1910), in Beijingshi danganguan, ed., *Beijing zilaishui gongsi dangan shiliao* (Beijing: Beijing yanshan chubanshe, 1986), 61. (京師自來水公司告白 1910年1月，北京市檔案館等編，北京自來水公司檔案史料 北京：北京燕山出版社，1986 年，第 61 頁。)

of winter to the torrid summer heat and whipping sandstorms, even if there were no customers.[53] The Waterworks initially planned to recruit seven hundred keepers, but the majority of potential applicants withdrew because of the rigid regulations and harsh penalties concerning attendance.[54] This shortage of hydrant keepers kept customers waiting in long lines to purchase running water. On top of that, the keepers seemed to have difficulties complying with modern regulations. For example, keeper Sun was absent without leave; while another keeper, Yang, was over an hour late. Both were fined a colossal half yuan as punishment.[55] The second problem was that since the Waterworks sold water mainly through public hydrants in the streets, the residents had to fetch the water themselves. Thus, many households were reluctant to switch to running water; they preferred to rely on water carriers who provided cheaper water with home delivery service. To a large extent, while running water in the public hydrants was a new source of drinking water, it did not address the last crucial step of water delivery to the doorstep.

The inability to provide in-house water was manipulated by water carriers to force the Waterworks to seek cooperation with them in piped communities where the proportion of homes with an internal water supply was less than 10 percent and the rest of the households either used well water or running water from the Waterworks' standpipes in the streets. To assuage the pressure from water carriers' resistance and also to promote running water, the Waterworks decided to hire existing water carriers to watch over open hydrants and to sell running water in the well houses' distribution territories that had not been completely superseded by piped communities. If anyone wanted to purchase running water, they just needed to buy water tickets from the Waterworks and then exchange the tickets for water from water carriers.[56]

Thus, water carriers extended their running water regime based on to the locations of the Waterworks and its branches. The traditional water carriers defended their existing regime and became the carriers for the new running water. In neighborhoods equipped with hydrants, water carriers began to dispense running water in their distribution territories. For example, Li San, a Shandong water carrier, owned a water distribution territory comprised of

53 *Shuntien Times*, February 17, 1910. (順天時報 1910 年 2 月 17 日 第七版 自來水屋。)

54 *Shuntien Times*, January 17, 1909. (順天時報 1910 年 1 月 17 日 第 6 版 京師近事。)

55 Beijing Municipal Archives, *Kanshou longtou shuifu weigui an*, J067-1-194, August 19, 1910. (北京市檔案館藏，J067-1-194，看守龍頭水夫違規案，1910 年 8 月 19 日。)

56 *Shuntien Times*, February 17, 1910. (順天時報 1910 年 2 月 17 日 第七版 自來水屋。)

FIGURE 9 A carrier of running water from street faucets, 1917–1919
SOURCE: SYDNEY D. GAMBLE PHOTOGRAPHS COLLECTION, DUKE UNIVERSITY LIBRARY

more than thirty households. He switched from well water to running water in 1910 and began distributing water for the Waterworks in his territory.[57]

With the end of the Qing dynasty, the cost of running water installations was not reduced. This situation is exemplified by Lü Xun (魯迅, 1881–1936), an eminent writer in modern China who in 1919 moved his family to a large compound in Beijing. He spent a total of 115 *yuan* to install running water. This amount was affordable for Lü Xun, whose monthly income was 300 *yuan*, while the average monthly salary of a primary school teacher in Beijing was about 40 *yuan*.[58] Along with the costly installation, running water bills were paid by the month rather than by the amount of actual consumption. Thus, several households would share one faucet and this caused deficits for the

57 Beijing Municipal Archives, *Lizhongshi bingkong gongxing chengzu shuidao weiyuan an*, J181-19-13863, April 1916. (北京市檔案館：爲報告事竊於本月二十七日奉派調查李鐘氏稟控龔姓承租水道違約把持一案，J181-19-13863，1916 年 4 月。)

58 *Lü Xun Riji* (Beijing: Renmin wenxue chubanshe, 1976), 322–323. (《魯迅日記》上卷 北京：人民文學出版社，1976 年，第 322–23 頁。)

Waterworks. Soon after, the Waterworks required piped residences to install water meters as the basis for water charges at a cost of 30 *yuan,* or to lease a meter from the Waterworks at eighty cents per month.[59] Because the installation of internal pipes and water meters was so expensive, the Waterworks relied on providing street hydrants with running water for the majority of its users and thus, as one newspaper observed, the prosperity of running water relied on water carriers. In return, the trusted water carriers used this advantage to monopolize running water.[60]

According to the regulations for running water, customers bought water from water carriers with prepaid tickets from the Waterworks. But water carriers would sell water without requesting tickets and would ask for money instead, as a common strategy for deliberately sabotaging the Waterworks. Individual carriers also used tickets as a ploy to increase their own earnings. Sometimes water carriers even stole tickets and resold them. For example, water carrier Li, who pocketed five tickets, was one of only a few who were caught doing this by the Waterworks inspector.[61] To prevent consistent fraud of this kind, the Waterworks installed water meters at all street hydrants as a means to supervise water carriers and monitor the volume of water consumption.[62]

Water carriers also used summer heat and drought to serve their ends. As water usage rose in the hot weather, the water carriers would increase water prices. Although the water carriers' actions contravened the Waterworks' regulations, the Waterworks turned a blind eye to prevent a boycott by the water carriers. In 1917 the summer drought dried up the Sun River and caused a shortage of running water, forcing the Waterworks to reduce the hours of business for hydrants from seven hours to five hours per day.[63] The water carriers in Xidan then increased water prices. If the Waterworks tried to oppose this, the water carriers would refuse to distribute running water. The Waterworks was fully aware of the shortcomings of its arrangement with the water carriers but

59 *Shuntien Times*, January 19, 1921. (順天時報 1921 年 1 月 19 日 第七版 自來水公司限制用水。) The mandatory installation of water meters led to widespread opposition and the requirement was suspended; see *Aiguo baihuabao,* May 13, 1921. (愛國白話報 1921 年 5 月 13 號 第四版 飭查撤銷水表。)

60 *Shuntien Times,* August 7, 1916. (順天時報 1917 年 8 月 7 日 第三版 自來水之弊端。)

61 Beijing Municipal Archives, *Longtou shuifu an*, J067-1-194, December 3, 1917. (北京市檔案館：龍頭水夫案，J067-1-194，1917 年 12 月 3 日。)

62 *Shuntien Times*, September 19, 1917. (順天時報 1917 年 9 月 19 日 第七版 龍頭安設水表。)

63 *Yishi baihuabao*, June 12, 1917. (益世白話報 1917 年 6 月 12 日 第三版 公司體恤管役。)

had no solutions.[64] The water carriers manipulated running water as their exclusive right in their territories, just as they had done with well water.

4 The Foreign Well Water Regime

With the advent of running water in 1910, more water carriers patronized Japanese well-drilling technology and foreign wells became the most popular type in well houses; the water in the old wells was so brackish that the water carriers couldn't compete with the better quality of running water. A police survey in Dongsi indicated that twelve out of seventeen foreign wells were drilled by water carriers following the establishment of running water.[65] By the end of the Qing dynasty in 1911, the number of water carriers engaging in the foreign well-water business had reached more than one hundred.[66] With the addition of these foreign wells, the municipal government of the newborn Republic ended the spatial inequality between the Inner City and the Outer City and drilled wells in both cities after the 1911 Revolution. The Metropolitan police, as representatives of the state, oversaw these wells for the purposes of sanitation and firefighting. By the mid-1920s the police owned and supervised at least four hundred and eighty out of three thousand public and private wells in Beijing.[67] Among the private wells, over 50 were registered at the Police Office, which allowed their owners to sell the water; and they were subject to a tax of $0.5 a month, while the rest of the water was retrieved from the public wells.[68] Although these public wells were used for public services, the police became a key actor in the water business and subleased suitable wells to water carriers in order to raise funds. For example, the head of police in the Outer Second Right District (外右二區) drilled a foreign well and sold water in the neighborhood to raise money for a welfare school.[69] The police department of the Outer Fifth

64 *Chenzhong bao*, June 28, 1917. (晨鐘報 1917 年 6 月 28 日 第五版 送水夫高擡水價。)

65 Beijing Municipal Archives, *The Survey of Foreign Wells in Beijing*, J067-001-00438, April 1917. (北京市檔案館檔案：北京洋井調查-調查東四牌樓以南街市洋井情形具結清單，J067-001-00438 ，1917 年 4 月。)

66 Wu Tingxie, ed., *Manuscripts of Beijing Gazetteer* (Beijing: Yanshan chubanshe, 1998), 468–469. (吳廷燮：北京市誌稿·貨殖誌 北京：北京燕山出版社，1998 年。)

67 "Beijing jingye diaocha," *Xinghua* 23, no. 32 (1926): 24. (北京井業調查，興華, 1926, 23 (32), 第 24 頁。)

68 "Water Supply: The Old and the New in China," *South China Morning Post*, July 29, 1926, p. 6.

69 *Jinghua ribao*, October 15, 1918. (京話日報 1918 年 10 月 15 日 第三期 甘泉助學。)

Right District (外左五區) planned to open four foreign wells and sell water to subsidize half-day schools.[70]

Water carriers would rent governmental wells run by the police to supplant their old wells and meanwhile build connections with the police. For example, water carrier Zhao sold water from a governmental well located at Dongan Gate (東安門).[71] Another water carrier rented a well located west of Caishikou (菜市口) that was owned by the police department.[72] Water carriers with foreign wells continued to be the principal competitors to running water, even though Capital Waterworks constantly appealed to the Ministry of Civil Affairs to fill in and shut down the drinking water wells in the city.[73] The Waterworks later petitioned for all wells with faucets to be replaced, and for the territory between water carriers and running water to be partitioned.[74] However, this was impractical since the coverage of water pipes was uneven and many districts had no access to running water, especially in the western section of the city that was farthest from Capital Waterworks in eastern Beijing. Moreover, the destitute municipal government itself was involved in the water carriers' business.

Water carriers used the governmental wells to shield themselves from running water's competition, but another of the reasons why they felt it necessary to keep their wells was the unreliability of running water. Compared to well water, running water that used rivers as a source was susceptible to drought and freezing. In June 1917 the source of running water was almost exhausted because of drought; the street hydrants were only able to supply water for less than one hour per day. One newspaper reflected that it was fortunate the wells had not been filled in, as the Waterworks had proposed in 1910.[75] The other factor that justified keeping the wells was that the street hydrants were prone to freezing in frigid weather. In the winter, Beijing's temperature would drop to −20°C and the outdoor faucets would freeze overnight, resulting in no water

---

70 *Chenbao*, July 1, 1920. (晨報 1920 年 7 月 1 日 第六版 開鑿水井補助學校。)

71 Beijing Municipal Archives, *Tu Heyun goujie bingding ouda shuifu an*, J181-019-02311, March 1913. (北京市檔案館：內左二區表送解杜赫雲勾結兵丁毆打水夫案卷，J181-019-02311，1913 年 3 月。)

72 *Shuntien Times*, December 28, 1927. (順天時報 1927 年 12 月 28 日第七版 承租官井售水之糾葛。)

73 *Qianshuo huabao*, June 17, 1911. (淺說畫報 1911 年 6 月 17 日 銷自來水的好辦法。)

74 "Nonggongshangbu wei zilaishui gongsi kuochong zhuanguan bing guandu shangban xiaoshi zhi minzhengbu ziwen," (August 9, 1911), *Lishi dangan*, no. 2 (1992): 64. (農工商部爲自來水公司擴充專管並官督商銷事致民政部咨文 宣統三年八月初九日，中國第一歷史檔案館, 清末北京擴充自來水專管史料, 歷史檔案, 1992 年第2期，第 64 頁。)

75 *Yishi baihuabao*, June 15, 1917. (益世白話報 1917 年 6 月 15 日 第二版 自來水要絕源。)

available the next morning and sometimes for the entire day.[76] During this period, water carriers would deliver and sell water from their own wells.

The constant breaching of pipes also contributed to the insecurity of Beijing's running water and reinforced the importance of the wells. In September 1913, water flowing from a cracked pipe quickly inundated the west street of Dashilai (大柵欄). As a local newspaper noted, Beijing constantly suffered from the fracture of running water pipes.[77] In June 1914, the pipe in Dashilai ruptured again and within minutes water was streaming through the street.[78] Pipeline fractures appeared to be common, with inhabitants complaining that frequent inundations from the conduit were damaging their houses and flooding the streets. The frequent breaks in pipes not only spilled hundreds of gallons of water but would temporarily cut off the water supply. An extreme example involved the running water pipe installed at Yamen of Nine Gates Infantry Commander (步兵統領衙門). On June 17, 1921, water leaking from a cracked pipe flooded Yamen. The chief commander was so annoyed with the frequent ruptures that he asked the Waterworks to dismantle the water pipes immediately, and he drilled four foreign wells to supply drinking water to Yamen as an alternative.[79] Nor is it surprising that Peking Union Medical College, a leading hospital and medical training center in China modeled after Johns Hopkins, built its own water supply system in 1917 because the supply from the Waterworks was inadequate and expensive.[80]

In spite of delivering running water, the water carriers never abandoned their own well water; they carried out the distribution of running water and well water simultaneously in the piped communities. For water carriers, the importance of retaining wells had more to do with their use as an alternative water source; the wells also became a strategy for counterbalancing running water. When dispensing water, it was not rare for water carriers to blend expensive running water with cheap well water. At the outset, the water carriers deliberately blended running water with dirty well water to undermine the reputation of running water because they were afraid the growing prosperity of running water would endanger their business and ultimately cast them aside.[81] One resident discovered this when he used running water to make tea

---

76 *Shuntien Times*, December 24, 1917. (順天時報 1917 年 12 月 4 日 第三版 保護自來水管。)

77 *Baihua jiebao*, September 19, 1913. (白話捷報 1913 年 9 月 19 日 五版 水管裂啦。)

78 *Aiguo baohuabao*, July 1, 1914. (愛國白話報 1914 年 7 月 1 日 四版 自來水管破裂。)

79 *Aiguo baihuabao*, June 19, 1921. (愛國白話報 1921 年 6 月 19 日 第四版 飭撤自來水管。)

80 RF: 04/001.02-070#1670: *Peking Union Medical College-Water Supply System*, Rockefeller Archive Center, New York.

81 *Shenbao*, September 13, 1910. (申報 1910 年 9 月 13 日，第 5，6 版 京師近事。)

and the tea turned exceedingly oily.[82]After the water carriers were in control of the distribution of running water, they began to manipulate blended water to maximize their profits.[83] Some water carriers even disguised themselves as representatives of the Waterworks while selling their well water. A well-house owner located at Dayuanjing Hutong (達圓鏡胡同) who employed seven water carriers put up a sign reading "Beijing Running Water" outside his well house and each water wheelbarrow was emblazoned with a sign reading "Beijing Waterworks." However, the water he distributed was entirely well water, which the Waterworks discovered and reported to the police.[84]

The water carriers also maneuvered the shortage of running water to take revenge on households with internal faucets. One resident named Zhu, who had running water in his house, found his tap water was out of order and he was without water for the entire day. His cook tried to buy two buckets of drinking water from a water carrier who was distributing water in the street. However, the water carrier refused on the grounds that Zhu's house used running water.[85] Another example from the community of Xihuamen (西華門) further illustrates the retaliation of water carriers. A public faucet was installed in the street near Xihuamen and residents converted it to use running water. Unfortunately, the Waterworks later shut the faucet down because of the cost. The water carriers reacquired the water supply and tripled water prices in retaliation. In 1920 these former running-water customers were suffering headaches from constantly using the salty water dispensed by water carriers, so they petitioned the Waterworks to reopen the former street faucet. This proved to be a futile attempt. The water carriers continued to distribute bitter well water and increased the price. If questioned, they suspended the water supply as revenge. Eventually one resident named Zhao could not stand this mistreatment and hired a person to carry water from other wells, but he was immediately stopped by water carriers who even broke into Zhao's courtyard and beat him in punishment.[86] Water carriers also used the opportunities created by

82 "Waicheng xunjing zongting wei yanna pohuai shuizhi zhitu dengshi zhi zilaishui gongsi yiwen" [外城巡警总厅为严拿破坏水质之徒等事致自来水公司移文], (August 19, 1910), in Beijingshi danganguan, ed., *Beijing zilaishui gongsi dangan shiliao*, 75.

83 *Shuntien Times*, June 18, 1918. (順天時報 1918 年 6 月 18 日 第九版 水夫以汙水供人飲料。)

84 Beijing Municipal Archives, *Jingshi zilaishui zonggongsi*, J067-1-225, December 15, 1918. (北京市檔案館：京師自來水總公司，J067-1-225，1918 年 12 月 15 日。)

85 *Shuntien Times*, November 8, 1921. (順天時報 1921 年 11 月 8 日 第七版 致北京自來水公司函。)

86 *Shuntien Times*, September 13, 1920. (順天時報 1920 年 9 月 13 日 第三版 挑水夫亦專制。)

the problems of running water, such as supply conditions, water charges, and technical problems, to make money. For instance, mechanical problems forced the shutdown of all the faucets in the city in August 1917, and water carriers did good business.[87]

The costly installation and high rates limited residents' ability to pay for running water. In the early twentieth century, outside of China, running water became a public good and was maintained by municipal governments to reduce the price; but Capital Waterworks in Beijing was run by a private company for profit. For example, the growing price of running water compelled residents who originally chose running water over wells to turn to water carriers again during the 1910s. In 1910, the price of running water was one copper coin for eight buckets of water; this price doubled in 1914.[88] Three years later the price doubled again to one copper coin for two buckets.[89] The price of running water increased fourfold between 1910 and 1920.[90] As one resident complained, running water was as precious as sesame oil; he could only afford to use it for making tea and not for cooking gruel.[91] These dramatic price increases forced customers to switch to water carriers, who also raised the price of well water. In general, Capital Waterworks was more interested in providing its product to reliable and affluent customers instead of expanding supplies and services to a larger population. According to a 1917 police survey, the southern police station outside Xuanwu Gate had thirty-six dwellings with plumbing, while the remaining one hundred families were supplied by water carriers from twelve private wells and three governmental wells.[92]

The water carriers successfully used running water's impediments to reinforce their foreign well water regime. According to Gamble's survey of the Dengshikou (燈市口) District, a community with 1,509 houses and 7,946 residents, the water supply for the district came either from ten wells or from the water mains of Capital Waterworks. Since plumbing was expensive and one copper bought only ten gallons of water, not many houses were supplied with running water. Therefore, most of the water had to be carried in buckets or wheelbarrows from the wells or street hydrants, and the people bought their

87 *Chenzhong bao*, August 9, 1917. (晨鐘報 1917 年 8 月 9 號 第五版 自來水總機傷損。)

88 *Aiguo baihuabao*, April 30, 1914. (愛國白話報 1914 年 4 月 30 號 自來水復原價。)

89 *Shuntien Times*, June 30, 1917. (順天時報 1917 年 6 月 30 日 第七版 水夫齊行漲價。)

90 *Shuntien Times*, January 25, 1921. (順天時報 1921 年 1 月 25 日 第七版 北京市民反抗自來水公司。)

91 *Shuntien Times*, November 12, 1920. (順天時報1920年十一月十二日 第七版 敬告北京自來水公司。)

92 Beijing Municipal Archives, *Nanzuo fenju yangjing qingxing jujie*, J067-001-00438, April 1917. (北京市檔案館：南左分局界內洋井情形具結，J067-001-00438, 1917 年 4 月。)

water by the bucketful from six well houses.[93] Despite the presence of running water, the water carriers served a surprisingly high 80 percent of the population in the city in 1922.[94] As a consequence, the future of running water was destined to be bleak because of its high cost and water carriers' resistance as well. In 1912 Capital Waterworks was already in deficit since costs were high while sales were limited. In order to save on expenditures, the Waterworks decided to shut down street hydrants with poor sales.[95] Until 1914, the Waterworks experienced even more financial deficits and planned to sell to the German company based in Tianjin that had constructed Capital Waterworks.[96] In 1918 the manager of the Waterworks, Wang Xitong (王錫彤, 1865–1938), felt ashamed about the meager profits and tried to resign.[97]

5 The Challenges of the Water Regime

The water carriers built a running water regime in piped areas and a foreign well water regime in unpiped areas of the city in order to adapt to the crisis brought by running water, but the presence of running water nonetheless added new tensions to the carriers' water regime. Disputes over water distribution territories were a chronic problem throughout the Qing period and the uneven distribution of running water created inequality among the water carriers. Within communities, water carriers who were closer to hydrants were more willing to collaborate with the Waterworks, while those who were farther away tended to favor selling water from foreign wells. The friction between running-water carriers and foreign-well-water carriers was inevitable. For example, carrier Jiang sold running water in Zhang's distribution territory. When caught, Jiang refused to give Zhang compensation and was beaten by Zhang's kin. Later on, he poured kerosene into Zhang's well and was sentenced to forty days' imprisonment.[98]

---

93 Sidney D. Gamble, *Peking: A Social Survey* (New York: George H. Doran Company, 1921), 325–29.

94 John B. Grant, *Report on a General Health Survey of Peking, China,* February 1922, RF RG 5 Series 2, SS- 601, Box 55, F346, Rockefeller Archive Center, New York, 30–31.

95 *Shuntien Times*, April 23, 1912. (順天時報 1912 年 4 月 23 日 第五版 酌閉龍頭。)

96 *Shuntien Times*, March 29, 1914. (順天時報 1914 年 3 月 29 日 第八版 賣自來水 。)

97 Wang Xitong, *Yizhai zishu* (Zhouzhou: Henan daxue chubanshe, 2001), 254. (王錫彤，抑齋自述 鄭州：河南大學出版社，2001 年，第 254 頁。)

98 Beijing Municipal Archives, *Zhang Qingyou konggao Jiang En Qingdao meiyou an*, J181-19-3228, April 1, 1913. (北京市檔案館：內右三區警察署張慶有控告姜恩傾倒煤油案，J181-19-3228，1913 年 4月 1 日。)

5.1 *The Internal Challenges*

The water revolution that stemmed from running water also brought in new water carriers from places outside Shandong Province, such as Zhili Province and Beijing.[99] For example, water carrier Liu, who distributed water at Babao Hutong (八寶胡同), was from Baoding (保定), the capital of Zhili Province.[100] The influx of new carriers challenged the dominance of Shandong carriers who had established themselves in the city's water business since the early Qing period. Accordingly, tension among carriers from different native places was common. Water carrier Yang from Ji County (冀縣) of Zhili Province delivered water in west Zhushikou (西珠市口), one of the most prosperous commercial districts in Beijing, and his business was extremely promising. However, his territory was encroached on by water carrier Liu, and in revenge, Yang jumped into Liu's well and contaminated the water after arguing with him.[101]

The water carriers usually resorted to violence in these conflicts. When water carriers Wei and Xie had a fight over the boundaries of their water distribution territories, Xie tried to kill Wei with a knife. The frightened Wei found shelter in a church but the deranged Xie committed self-mutilation by cutting his own head.[102] In the 1920s these disputes resulted in even more ruthless violence. Water carrier Du had a dispute with carrier Fu over the rights to distribute water to one household. Du convened forty carriers while Fu summoned thirty carriers for a mass brawl, all carrying poles as weapons. The fray concluded with twelve broken poles, seven serious injuries, and eleven minor injuries.[103] An even more brutal example involved water carrier Lang outside of Qianmen. As he was delivering water, carrier Zhou accused him of trespassing on his territory and overturned Lang's wheelbarrow. Lang fought back and was struck on the face by Zhou's shoulder pole. The blow was so hard that Liang's right eyeball flew out at the scene.[104]

This water violence within the water regime caused injuries and even death, which led to the imprisonment of the involved water carriers and undermined their businesses. Accordingly and gradually, more water carriers resorted to

---

99 Li Youying, "Survey of Laborers in Beijing," *New Youth* 7, no. 6 (1920): 14–17. (李幽影：《北京劳动状况》，《新青年》第 7 期第 6 号，1920 年 5 月，第 14–17 页。)

100 Beijing Municipal Archives, *Cui Erxiang qie shuigou masheng an*, J181-019-11148, May 30, 1916. (北京市檔案館：內左一區表送崔二祥等咎竊水鉤麻繩一案, J181-019-11148, 1916 年 5 月 30 日。)

101 *Chenbao*, July 30, 1923. (晨報 1923 年 7 月 30 日 第六版 水夫爭道投井。)

102 *Aiguo baihuabao*, September 4, 1913. (愛國白話報 1913 年 9 月 4 號 四版 爭水道拿刀動兇。)

103 *Chenbao*, June 29, 1923. (晨報 1923 年 6 月 29 日 第六版 挑水夫之一場血戰。)

104 *Chenbao*, October 4, 1924. (晨報 1924 年 10 月 4 日 第六版 水夫打丟眼球。)

comparatively peaceful maneuvers—seizures of water wheelbarrows became a common practice in territory conflicts. The confiscation of wheelbarrows would interrupt water distribution and cause losses for opponents, which was more effective than physically assaulting carriers. For example, well-house owner Yao sent one of his carriers to deliver water to resident Huang's house. Water carrier Zhang seized Yao's water wheelbarrow and accused Yao of encroaching on his distribution territory.[105] When well-house owners Wei and An had a dispute over two customers, Wei seized two of An's water wheelbarrows by force.[106] Another carrier Zhang had a dispute over territory with Lu and seized Lu's water wheelbarrow. Lu wrote an anonymous letter to the police, falsely accusing Zhang of concealing firearms and ammunition.[107] In addition, water carriers would sabotage their rivals' businesses by contaminating their wells. Water carrier Tian drilled a new foreign well and sold water; the water was very popular and his business was thriving. However, his rival poured kerosene into the well one night, and Tian had to spend extra money to clean his well.[108] To conclude, running water complicated the city's relationship to water and changed the dynamic of the water regime. Meanwhile, the changing nature of the street—including the increasing rate of suicides, growing crime, new traffic modes, and the presence of warlords—imposed additional new challenges to the water regime.

5.2 The Changing Nature of the Street

In the Beiyang period, Beijing was famous not only for its venal politicians, rapacious warlords, and idealistic students, but also for its courteous but insistent policemen, rancorous mule drivers, and night-soil carriers.[109] Water carriers' street confrontations involved suicides, theft, traffic, soldiers, and other professional people who shared the streets. Their first challenge was well protection. As the source of drinking water, wells were essential to the water

105 Beijing Municipal Archives, *Waizuo yiqu biaojie Hao Lixuan kong Zhang Zihe jiangyi shuiche tuizou yian*, J181-019-00022, July 20, 1913. (北京市檔案館：外左一區表解郝立軒控張子和將伊水車推走一案，J181-019-00022，1913 年 7 月 20 日。)

106 Beijing Municipal Archives, *Neiyou sanqu biaosong An Zitian kong Wei Dianchen qinzhan shuiche yian*, J181-019-21909, February 9, 1918. (北京市檔案館：內右三區表送安子田等控魏殿臣侵占水車一案，J181-019-21909，1918 年 2 月 9 日。)

107 Beijing Municipal Archives, *Zhang Zengshou yu Lu Taigui yin zheng shuidao zhengchao yian*, J181-19-55781, December 1927. (北京市檔案館：張增壽與路臺桂因爭水道爭吵一案，J181-19-55781，1927 年 12 月。)

108 *Shishi baihuabao*, December 22, 1922. (實事白話報 1922 年 12 月 22 日 楞往井內倒煤油。)

109 David Strand, *Rickshaw Beijing: City People and Politics in the 1920s* (Berkeley: University of California Press, 1989), 15.

business. If any impurity should find its way into a well and contaminate its water, the water carriers would have to remove the offensive substance. It was in the interests of the water carriers to supply their customers with pure water; thus it was to their advantage to carefully guard their wells against pollution. However, wells in the streets were vulnerable to contamination. When well owner Liu was drawing water from his well, an idle man who seemed to suffer from a mental disorder threw a red pill into the well, forcing Liu to shut down the well for a period of time.[110] Some wells were even polluted deliberately. Two wells, in Xueyuan Hutong (學院胡同) and Taipingqiao (太平橋), respectively, had kerosene poured into them, contaminating the water.[111]

The most severe problems of well contamination came from suicides. Situated in the streets with easy accessibility, these wells were the most convenient place to commit suicide. Suicides by the indigent accounted for one of the highest rates of suicide in Beijing. This was especially true for suicides in poverty-stricken banner families. The 1911 Revolution ended the preferential treatment that the banner people had previously enjoyed, including the gradual reduction of stipends, which caused banner families to fall into financial crisis and contributed to the increasing suicide rate.[112] In 1914 a young bannerman named Yu drowned himself in a well because of impoverishment.[113] In 1918 a man who was a destitute morphine addict jumped into a foreign well located in north Dongchang Hutong (北東廠胡同); although the well owner immediately rescued him and prevented the well from being contaminated.[114] In 1922 an aged man who was mistreated by his son and daughter-in-law tried to commit suicide in a well situated in Jinger Hutong (井兒胡同). Fortunately, the well owner Liu observed this and saved him.[115]

Among all the suicides, women's suicides in particular were a significant source of potential water pollution. A despairing widow drowned herself in the well in the courtyard of a family named Ma (馬家大院) when she noticed that the owner had gone out.[116] A young wife entered a well house on the pretext of

110 Beijing Municipal Archives, *Liu Bozhen kong Li Deqi wang shuijingnei touqi hongwan*, J181-19-47781, April 30, 1925. (北京市檔案館：劉伯珍控李德啓往水井內投棄紅丸案，J181-19-47781，1925 年4 月 30 日。)

111 *Aiguo baihuabao*, January 27, 1919. (愛國白話報 1919 年 1 月 27 號 第四版。)

112 Janet Chen, *Guilty of Indigence: The Urban Poor in China, 1900–1953* (Princeton, NJ: Princeton University Press, 2012).

113 *Aiguo baihuabao*, August 13, 1914. (愛國白話報 1914 年 8 月 13 號 三版 男子輕生。)

114 *Jinghua ribao*, August 1, 1918. (京話日報 1918 年 8 月 1 日 第三版 不紮就死。)

115 *Beijing baihuabao*, May 30, 1922. (北京白話報 1922 年 5 月 30 號 第二版 忤逆子逼父投井。)

116 *Chenbao*, July 30, 1922. (晨報 1922 年 7 月 13 日 第七版 孀婦因何棄女投井。)

getting drinking water and jumped into the well in Shenlujie (神路街) because of domestic violence. The panicked water carrier called the police and rescued the woman. In a lawsuit, the water carrier demanded that the woman's husband clean the well.[117] A well in Youanmen was a noted sweet-water well and residents relied on it. One morning the water carrier found a dead body in the well, which turned out to be a young woman from the neighborhood suffering from a mental disorder.[118] Concerned about the increasing number of suicides in the wells, the owners adopted protective measures such as locking the wells with covers at night and building houses over the wells, but these precautions could not ward off unexpected suicides. Well-house owner Zhang had a foreign well in Dongsi. One morning an unknown man approached him on the excuse of looking for an acquaintance who worked in Zhang's well house and asked for a drink of water. With a borrowed bowl in hand, the man jumped into the well. The astonished Zhang climbed down into the well and retrieved the man from the water.[119]

A second challenge for the water regime arose from theft. Increasing impoverishment and social unrest led to growing crime in Beijing, and the well houses located in the public streets were susceptible to theft. For example, a thief broke into one well house situated in east Zongbu Hutong (总布胡同) when the carriers were out distributing water. All their belongings, consisting of clothes, cash, and even bedding, were pilfered.[120] Devices such as water buckets, hooks, and ropes were prone to theft as well. When water carrier Zhang left his two buckets outside, they were stolen by a destitute migrant worker.[121] Water carrier Liu parked his water cart outside the well house and later found two iron hooks on it had been taken away.[122] Water carrier Yan had his rope cut down by a morphine addict.[123] Water carrier Liu had his winter coat stolen from the clothesline in his yard.[124] The penniless bannerman Wen,

---

117 *Chenbao*, May 6, 1925. (晨報 1925 年 5 月 6 日 第六版 少婦氣憤投井。)

118 *Cheobao,* November 27, 1926. (晨報 1926 年 11 月 27 日 第六版 郭家井內女屍。)

119 Beijing Municipal Archives, *Beipingshi jingchaju zuosiqu guanyu Zhang Wu yin heshui toujign an*, J183-002-02287, July 1926. (北京市檔案館：北平市警察局左四區關於張武因喝水投井案，J183-002-02287，1926 年 7 月。)

120 *Aiguo baihuabao*, October 17, 1913. (愛國白話報 1913 年 10 月 17 日 第四版 水屋被竊。)

121 *Shijie ribao*, November 3, 1927. (世界日報 1927 年 11 月 3 日 第七版 偷水桶的賊。)

122 Beijing Municipal Archives, *Neizuo yiqu biaosong Cui Erxiang jiuqie shuigou masheng yian*, J181-019-11148, May 30, 1916. (北京市檔案館：內左一區表送崔二祥等咎竊水鉤麻繩一案，J181-019-11148, 1916 年 5 月 30 日。)

123 Beijing Municipal Archives, *Chenjie Liu Shuyin touqie Yan Futang jingsheng*, J181-19-43097, January 7, 1925. (北京市檔案館：呈解劉樹寅偷竊閻福堂井繩，J181-19-43097，1925 年 1 月 7 日。)

124 *Shijie ribao*, October 3, 1927. (世界日報 1927 年 10 月 3 日 第七版 賊人拋贓脫逃。)

who made a living by collecting wastepaper in the street, stole one rope and two hooks from three well houses to exchange for money.[125]

The third challenge derived from the changing mode of the street traffic with the arrival of the automobile. In the early twentieth century, Beijing had a modern tram service and cars in its streets. Accidents between cars and water carriers, who pushed heavy wheelbarrows and delivered water in the streets, were not uncommon. Water carrier Zhang, distributing water during a snowy day, was hit by a car. He suffered only minor injuries, but his water cart was seriously damaged.[126] Water carrier Zheng had parked his water cart when a car suddenly ran into it. Zheng was struck and taken to the hospital with a broken leg and pelvis; the driver was taken into custody.[127] Water carrier Li was hit by a car driven by a Frenchman; his water cart was knocked over and his face injured. The driver tried to flee but was stopped by the police.[128] One reckless driver punched a water carrier because he felt that the water carrier was blocking his way; he then deliberately struck his own head in order to falsely accuse the water carrier.[129] As pedestrians, water carriers were usually the victims in auto accidents, but some water carriers fought back. Water carrier Zhou had his water cart rammed by a speeding car, which made him so angry that he pursued the hit-and-run driver on foot and eventually caught him.[130]

The fourth challenge for water carriers stemmed from the increasing presence of soldiers in the city. The capital was full of troops and soldiers in the warlord period (1916–1927).[131] The water carriers were often the targets of military violence. When well-house owner Zhao argued with a salesman in a flour store, he was beaten by a royal guardsman who had been befriended by the

125 Beijing Municipal Archives, *Jiesong xianzang zeifan Wen Shuangshou, Baorui deng erming*, J181-19-19030, September 3, 1918. (北京市檔案館：解送現贓賊犯文雙壽寶瑞等二名，J181-19-19030，1918 年 9 月 3 日。)

126 Beijing Municipal Archives, *Neiyou yiqu chengbao Guo Qinli jiaqiche pengshang tiaoshuifu Zhang Keqing, sunhui shuiche an,* J181-18-10406, January 1919. (北京市檔案館：內右一區呈報郭琴利駕汽車碰傷挑水夫張克慶損毀水車辦理情形，J181-18-10406，1919 年 1 月。)

127 *Chenbao*, October 17, 1924. (晨報 1924 年 10 月 7 日 第六版 汽車撞傷水夫。)

128 *Aiguo baihuabao*, June 17, 1921. (愛國白話報 1921 年 6 月 17 號 第四版 洋人汽車肇禍。)

129 *Beijing baihuabao*, February 15, 1921. (北京白話報 1921 年 2 月 15 日 第三版 司機真逞蠻橫。)

130 *Chenbao*, September 3, 1924. (晨報 1924 年 9 月 3 日 第六版 水夫一躍上汽車 抓住汽車夫算賬。)

131 The "warlord period" is a term used to describe the demise of central government authority and the existence of regional political entities controlled by military cliques in China from 1916 to 1927 after the death of President Yuan Shikai.

salesman.[132] When carrier Cheng was delivering water, a soldier named Lu who was playing soccer with a group of soldiers approached Cheng and asked for drinking water. Cheng did not hear Lu's request and kept moving. Lu then punched him in the face, and seven or eight soldiers joined in beating him, using his carrying pole.[133] Such violence was not limited to Chinese soldiers. A British soldier bumped into water carrier Niu as he was carrying water and broke his two buckets; when Niu went to argue with the soldier, who was hiding in a brothel, the soldier beat him.[134] In addition to the soldiers, rampaging military vehicles also presented a threat to water carriers in the streets. Water carrier Wang was carrying water in his cart and was struck by a military vehicle, damaging the cart. Wang, who suffered minor injuries, asked for twenty *yuan* in compensation, but the driver only agreed to pay five.[135]

In conjunction with these incidents, soldiers could also undermine the sanitation of well water. When a solider named Li lost his military uniform and identity papers while intoxicated, the panic-stricken solider tried to commit suicide in water carrier Zhang's well.[136] During the warlord period it was common to find abandoned ammunition in wells. Water carrier Wang found an abandoned gun in his well and reported it to the police.[137] Soldiers not only imposed external challenges to water carriers but were also involved in the water business. One army man, Xie, who served in the headquarters of the Municipal Garrison Forces, leased a water distribution territory in Tianqiao

132 Beijing Municipal Archives, *Neizuo erqu biaosong Du Heyun goujie bingding ouda shuifu anjuan*, J181-019-02311, March 1913. (北京市檔案館：內左二區表送解杜赫雲勾結兵丁毆打水夫案卷，J181-019-02311，1913 年 3 月。)

133 Beijing Municipal Archives, *Waiwuqu biaosong tuishuifu Cheng Zhanyuan yu junren Lv Disheng yin qiushuihe huxiang jiuda an*, J181-019-47273, May 1919. (北京市檔案館：外五區表送推水夫程占元與軍人呂第生因求水喝互相揪打各有微傷案，J181-019-47273, 1919 年 5 月。)

134 Beijing Municipal Archives, *Neizuo yiqu xiangbao banli yingbing oushang tiaoshuiren Niu Jiachen qingxing an*, J181-018-02616, June 20, 1914. (北京市檔案館：內左一區詳報辦理英兵毆傷挑水人牛甲臣情形案 J181-018-02616，1914 年 6 月 20 日。)

135 Beijing Municipal Archives, *Zhongyiqu biaosong disanjun qiche sijijunren Liu Haogu qiche pengshang tuishuiren Wang San yaobu yian*, J181-019-42765, November 9, 1924. (北京市檔案館：中一區表送第三軍汽車司機軍人劉好古汽車碰傷推水人王三腰部一案，J181-019-42765，1924 年 11 月 9 日。)

136 Beijing Municipal Archives, *Neizuo siqu biaosong junren Li Zhiwang zai Zhang Bin shuijing xunsi an*, J181-019-50474, July 1926. (北京市檔案館：內左四區表送軍人李誌旺在張斌水井尋死案，J181-019-50474，1926 年 7 月。)

137 Beijing Municipal Archives, *Neiyou siqu chenbao tiaoshuiren Wang Xizeng you shuijing laochu feiqiang yizhi*, J181-018-07913, October 1917. (北京市檔案館：內右四區呈報挑水人王錫增由水井內撈出廢槍一支，J181-018-07913，1917 年 10 月。)

(天橋) at the rate of one hundred *yuan* for four years and sold water.[138] Military power was also used in water disputes. The soldier Sun, who was a guardsman in the staff division of Beijing, helped his townsmen retake a well from a resident.[139]

The last challenge that water carriers faced came from other professional people working in the streets, such as street sprinklers, night-soil collectors, and horse-carriage drivers. The streets in Beijing were dusty throughout the dry season, so the government hired street sprinklers to spray water on the streets to keep the dust down. Two men would carry a large tub of water into the center of a street and then give it an even sprinkling by hand, using a willow basket on the end of a pole. The sprinklers used water from street wells and had frequent contact with water carriers in the streets, so conflict was inevitable. For example, street sprinkler Song spilled water on water carrier Jiao and wet his clothes. The two fought and both were detained by the police.[140] Horse-carriage drivers and horses usually quenched their thirst from water carriers' wells, which could also cause tensions. Water carrier Wang was distributing water on a hot afternoon when carriage driver Gao wanted some water. Wang rejected his request and as a result the two fought, leaving Wang with a bleeding head and Gao in a coma.[141]

6 Conclusion

Running water did not circulate throughout the urban fabric in a progressive fashion as one might have assumed; while the water carrier system in operation before the advent of centralized water was neither rudimentary nor inefficient. Beijing's running water was government-supervised and merchant-managed. The priority of the municipal government was to use free water from the centralized waterworks as a tool with which to fight fires more efficiently. Moreover, because of chronic financial stringency, the Beijing

138 Beijing Municipal Archives, *Waiyou erqu biaosong junren Xie Chunsheng yin shuiche jiuge yujiang Ning Xueben daizou yian*, J181-019-51032, October 1926. (北京市檔案館：外右二區分區表送軍人謝春生因水車糾葛欲將寧學本帶走等情一案，J181-019-51032, 1926 年 10 月。)

139 Beijing Municipal Archives, *Neiyou siqu biaosong Xun Zhaoshi kong Zou Lisi daitong junren siru minzhai qiangshou shuijing yian*, J181-019-55458, November 1927. (北京市檔案館：內右四區表送勛趙氏控鄒李氏帶同軍人私入民宅強收水井一案，J181-019-55458，1927 年 11 月。)

140 *Baihua jiebao*, October 3, 1913. (白話捷報 五版 1913 年 10 月 3 號 水門傷人。)

141 *Chenbao*, July 2, 1926. (晨報 1926 年月 2 日 第六版 倒水夫與車夫血戰。)

municipal government never tried to municipalize running water to improve the city's household water supply, as its western counterparts did. As a result, the abundance, permanence, and regularity of running water was artificially restricted because of Capital Waterworks' lack of funds and inadequate methods of distribution. The internal problems of Capital Waterworks, combined with pushback from the water carriers, forced the Waterworks to seek cooperation with the carriers who began dispensing running water in their water distribution territories. To a large extent, the traditional water carriers built a new water regime by using running water as a new source of drinking water and successfully competing within the differentiated markets for the control and sale of potable water. At the same time, water carriers maintained their well water regime in the unpiped areas.

Running water complicated the city's relationship to water. The 1911 Revolution collapsed the dichotomy in drinking water between the Inner City and the Outer City. But the water carriers were unable to build a unified and coherent water regime throughout the city because running water created a new spatial pattern by dividing the city into piped areas and unpiped areas. What's more, the running water brought in new water carriers and intensified the relations between carriers. The internal challenges along with the changing nature of the streets necessitated the citywide coordination of water carriers, it also justified the water carriers' solidarity and prepared carriers to be "water lords."

As the transformation of Beijing into a modern city with new spatial and mobility patterns continued, the water carriers were able to adapt to modern running water until 1923, when the aggressive expansion of Capital Waterworks destroyed the fragile balance between running water and the water carriers. Thus, the solidification of the water regime was not just a matter of practical exigency but also a pretext for the establishment of new forms of water governance that would transform the water carriers, who would then leverage the authority of their well guild to negotiate, regulate, and maintain order within the water regime, as well as redefine the water's relationships with the city.

CHAPTER 4

The Making of the Water Lords, 1924–1927

In 1925, a resident who witnessed a fight breaking out between water carriers and the Waterworks wrote to the local newspaper and called the water carriers in Beijing rude and unreasonable. He cited their insults to the police and their threats to bury workers alive. The letter writer observed that this was the Republican Period in which even the emperor had been expelled from the Forbidden City. How could the autocracy of water carriers be endured?[1] As early as 1924, Beijing residents had crafted the term "water lord" (水閥) from the term "warlord" (軍閥) and used it as a sobriquet for the water carriers to imply that they were now controlling and manipulating water distribution in specific areas of Beijing and had resorted to violence to maintain their control. In other words, every carrier had his own particular sphere of influence over water usage, and it was generally impossible for a consumer to purchase water from a carrier other than the carrier in whose beat he lived.[2]

The period of 1924 to 1927 is the height of the water carriers' power and is when the water lords are finally taking shape. The expansion of Capital Waterworks in 1924 smashed the delicate balance that had existed between it and the water carriers. The water carriers terminated their cooperation with the Waterworks and became more aggressive in resisting running water. The persistent internal and external challenges compelled the water carriers to reconstruct a citywide water regime. Accordingly, they reorganized the well guild to solidify their power and maneuvered political movements to articulate their interests. The water carriers were nonetheless able to maintain their regime in the face of all the state and police efforts to control them. To become water lords, the carriers manipulated the underfinanced municipal government's reluctance to provide clean water and the police participation in the well business. At the same time, the weakened collective power of the neighborhoods, arising from the elite families' switch to running water, further contributed to the making of the water lords.

Water carriers' increasing political sophistication and participation in the city culture of Beijing after the 1919 May Fourth Movement facilitated their

1 *Shuntien Times*, October 26, 1925. (順天時報 1925 年 10 月 26 日 第七版 水車夫之無理取鬧。)

2 *Shuntien Times*, April 29, 1924. (順天時報 1924 年 4 月 29 日 第七版 水閥糞閥之大壟斷市民束手叫苦。)

 | DOI:10.1163/9789004720800_006

response to the Waterworks. Throughout these tumultuous times from 1924 to 1927, the water carriers displayed prudence in charting a course for building their water regime despite the political uncertainties and social unrest. Meanwhile, the Beijing municipal government was not intimately involved with water distribution because drinking water was considered a private commodity rather than a public utility. The warlord government's only responsibility was to mediate between water carriers and the Waterworks, which meant that the government tolerated the water carriers' violence and induced the transformation of water carriers into water lords. In this respect, warlordism created an opportunity for those seeking to exploit the power vacuum, and the water carriers were only one of the non-state actors seeking to maximize their own benefit. The power of the water carriers was not curtailed until the coming of the Nationalist government in 1928.

1 Organizing the Carriers' Association

Because access was one factor that inhibited the consumption of running water, the Capital Waterworks moved to break free of its reliance on the uncontrollable water carriers in 1918 and began promoting consumption by creating its own team of carriers—who were independent from the established water carriers—to deliver running water directly to people's doorsteps.[3] The Waterworks also set about expanding its water network in 1919 by adding mains to the streets to meet the growing demand.[4] Shortly after, the Waterworks decided to purchase iron pipes to install in additional streets.[5] The Waterworks invited a Danish engineer to plan further expansion and contracted for pipes from the Siemens Company of Germany in 1923.[6] This gradual expansion of running water steadily encroached on the territories of water carriers and infringed on the previous cooperation in water delivery between the carriers and the Waterworks. To resist this encroachment, water carriers decided to build the well guild to defend their water rights.

The water carriers had an internal cohesion based on kinship and native ties to Shandong Province in the Qing period. In an arrangement that began in

3 *Chenzhongbao*, August 17, 1918. (晨鐘報 1918 年 8 月 17 日 第六版 自來水公司設所送水。)
4 *Shuntien Times*, December 24, 1919. (順天時報 1919 年 12 月 24 日 第七版 擴充自來水管。)
5 *Beijing baihuabao*, February 7, 1920. (北京白話報 1920 年 2 月 7 日 第三版。)
6 "Zilaishui gong ni zeyao luxu kuochong jihuashu," in Beijingshi danganguan, ed., *Beijing zilaishui gongsi dangan shiliao* (Beijing: Beijing yanshan chubanshe, 1986), 124–126. (《自來水公司擬擇要陸續擴充計劃書》，北京市檔案館等編：《北京自來水公司檔案史料》北京：北京燕山出版社, 1986年，第 124–26 頁。)

the seventeenth century to reward them for aiding the Manchus in conquering China, the state had granted Shandong men the right to pass the leaseholds of their well houses on to their sons. Consequently, the same Shandong families maintained control over wells in perpetuity and consistently hired water carriers who, like themselves, came from Shandong Province. Both well-house owners and water carriers in Beijing maintained these kinship connections and native-place ties to Shandong Province; thus, many families in Shandong Province were involved in the water business in Beijing. For example, the Wang family, originating from Rongcheng County (榮成縣) on the Shandong peninsula, owned a well house and had run a water business in Changxiang toutiao (長巷頭條) since the Qing period.[7] The Liang family, also from Rongcheng, had a water business and well in Western Xiejie (西斜街).[8]

Water carriers' ties to their native place of Shandong were the foundation of the water regime in both the Inner City and the Outer City. So strong were these linkages that Shandong water carriers continued to monopolize the trade throughout the Republican period. This is exemplified by the case of Zhu Donghai. Zhu came from Fushan County (福山) in Shandong Province and ran a water business in Beijing, following the family tradition. While the Palace had monopolized the famous Jade Spring since the Qing period, after the 1911 Revolution the spring was reopened to the public. In 1918, Zhu created a soft drink company called the Jade Spring Soft Drink Company and used pure Jade Spring water to make soda water. To show respect and also to popularize his water, Zhu presented ten boxes of soda water to the deposed emperor Puyi.[9] While this example illustrates the continuing role of well owners with Shandong connections in the Republican period, another case demonstrates that ordinary water carriers also maintained their Shandong connections. Zhang Xiuqing came to Beijing from Shandong as a teenager and carried water in Di'anmen (地安門). Zhang had two elder brothers. His eldest brother also carried water in Beijing, and his second brother tilled the land in their hometown. Zhang married in Beijing and sent his wife back to Shandong

7 Beijing Municipal Archives, *Shandong rongchengxian zhixian guanyu Wang Yuqian yin shuijing jiuge su Wang Dexiu yian*, J181-18-14880, January 1922. (北京市檔案館：山東榮成縣知事關於王毓乾因水井糾葛訴王德秀等所供各節請派警調查明確的函，J181-18-14880，1922 年 1 月。)

8 "Sun Shufa konggao Liang Zongxin yin shujing jiuge yian panic," in Xiong Yuangan, ed., *Jingshi defang shenpanting facaohui pandu huibian* (Beijing, 1914), 40. (孫樹發控告梁宗興因贖井糾葛一案判詞，熊元翰編：《京師地方審判廳法曹會判牘匯編》1914，第 40 頁。)

9 *Jinghua ribao*, June 16, 1918, and July 30, 1918. (京話日報 第四版 1918 年 6 月 16 日 呈進汽水; 京話日報 1918 年 7 月 30 日 玉泉山啤酒汽水公司開彩。)

after she had children.[10] Their Shandong roots created and reinforced kinship bonds between water carriers. Meanwhile, their native-place network generated connections (*guanxi*) that served to extend their contacts, create mutual obligations and protections, and establish social distance and hierarchy.[11]

Worship of the well god further consolidated this cohesion among carriers within the water regime. The cult of the well god was popular in northern China in the Late Imperial period, and the well god was used as the trade deity of the water business. According to Duara's notion of the "cultural nexus of power," religious symbols are capable of being molded for opposing goals, while retaining their conative power.[12] Water carriers created links between their occupation and religion by embedding the norms of the well deity into their networks, which allowed them to effectively access resources and exercise power. Each well house had a statue of the well god, and well owners offered sacrifices on the first and fifteenth day of the lunar calendar (see fig. 10). In some streets, water carriers built remarkable temples for the well god during the Qing period. One temple for the well god in the western portion of the city was also used as a dormitory to accommodate water carriers.[13]

In view of the growing competition with running water, Liu Tongliang, a water carrier who had come from Shandong Province, petitioned the Ministry of Internal Affairs in 1920 to reorganize the water carriers' guild that had been formed in the late Qing period under the pretext of improving their livelihoods.[14] However, the Ministry was deeply concerned that the proposed association would strengthen the carriers' monopoly on water distribution, so it instructed the Police Department to thoroughly investigate the application before approval.[15] Nonetheless, the Ministry's intention to thwart the well guild did not deter its reestablishment. In 1923 the water carriers finally revived their organization so that it could serve to mediate disputes and provide mutual assistance based on their Shandong connections. It even managed to establish

10 Beijing Municipal Archives, *Zhongyiqu guanyu tishu canyiyuan Tao Jingwu kong tiaoshuiren Zhang Xiuqing you touqie yifu xianyi yian*, J181-019-26862, September 12, 1920. (北京市檔案館：警察廳中一區關於提署參議員陶經武控挑水人張秀清有偷竊衣服嫌疑一案的呈，J181-019-26862，1920 年 9 月 12 日。)

11 S. A. Smith, *Like Cattle and Horses: Nationalism and Labor in Shanghai, 1895–1927* (Durham: Duke University Press, 2002), 20–21.

12 Prasenjit Duara, *Culture, Power, and the State: Rural North China, 1900–1942* (Stanford, CA: Stanford University Press), 15–41.

13 *Beijing Yishibao*, January 31, 1917. (北京益世報 1917 年 1 月 31 日 第三版 火神誠心起哄。)

14 *Shuntien Times*, March 23, 1920. (順天時報 1920 年 3 月 3 日 第十一版 水夫亦要立會。)

15 *Beijing Baihuabao*, April 2, 1920. (北京白話報 1920 年 4 月 2 日 第三版 調查內容。)

FIGURE 10 A well-god shrine behind the well, 1920–1921
SOURCE: WILDER HAYDN HAINES PAPERS, PRINCETON UNIVERSITY LIBRARY

a public cemetery for water carriers.[16] Although the guild was not officially recognized by the Ministry, its members demonstrated great solidarity about controlling the water supply in Beijing.

Internal disputes including tensions between owners and carriers, strains between carriers, and carriers' friction with residents in everyday business also justified the establishment of the well guild, a water organization to help alleviate tensions and adjudicate contests. In the 1920s, Beijing had 300 well-house owners and 2,200 water carriers.[17] The owners dominated the water business and were quite wealthy. In contrast, water carriers would sometimes die from the overwhelming stress of carrying water, especially from distant wells or in bad weather. For example, carrier Chen died at the age of forty-two from overwork.[18] Although the kinship ties and native-place connections among water carriers and owners were strong, tensions between them were not rare. A poverty-stricken carrier Zhang stole owner Yue's water wheelbarrow and tried to sell it to another owner, Liu, for ten *yuan*. Unfortunately, Liu was Yue's relative and recognized the stolen water cart. As a result, Zhang was seized and

16 *Chenbao*, November 15, 1923. (晨報 1923 年 11 月 15 日 第六版 擔水夫有了團體。)
17 Sidney D. Gamble, *Peking: A Social Survey* (New York: George H. Doran Company, 1921), 430.
18 *Chenbao*, August 25, 1922. (晨報 1922 年 8 月 25 日 第七版 水夫過勞致命。)

escorted to the police.[19] Well-house owner Wang accused carrier Fu of stealing his belongings and fired him; the annoyed Fu poured kerosene into Wang's well in revenge.[20]

In their daily operations, water carriers also intensified their relations with residents. For example, water carriers squeezed the residents at Beixinqiao (北新桥) with high water rates, and would suspend water delivery for three to five days if the price was rejected.[21] Such discord with residents would escalate into fights between carriers and residents. When carrier Song's water cart bumped his customer Wang's wheelbarrow, Wang enlisted his parents in a fight with Song and, together, they injured him.[22] Water carriers were also susceptible to suspicion because they went in and out of customers' households on a daily basis. For example, carrier Zhang was falsely accused of stealing clothes from a homeowner's chef.[23] The trivial disputes that arose from daily business further rationalized the formation of the well guild.

In February 1925, inspired by a nationwide campaign for workers' rights and union power in China, well-house owner Liang Chunhe reapplied for formal recognition of the well guild, but his request was again denied by the authorities. Even so, as an informal trade guild, the water carriers extended their hidden authority over the water business and were involved in water pricing, disputes between water carriers and residents, and internal conflicts among carriers. Its members were obligated to pay twenty coppers per month for the daily operations of the guild, which included the rental fee for the guild hall, salaries for the staff, and pensions.[24] In April 1925, the water carriers were concerned that the installation of public faucets would jeopardize their business, so Liang again tried to ally with water carriers in the city to reestablish a well

19 Beijing Municipal Archives, *Jingshi jingchating zongyiqu guanyu Zhang Anjia touqie shuiche cheng*, J181-019-27125, December 24, 1920. (北京市檔案館：京師警察廳中一區關於張安家偷竊水車的呈，J181-019-27125，1920 年 12 月 24 日。)

20 Beijing Municipal Archives, *Neizuo siqu biaosong junren Wang Wenliang kong Fu Hai wang yixiong Wang Huanzhang shuijingnei Qingdao meiyou yian*, J181-019-42140, October 1924. (北京市檔案館：內左四區表送軍人王文亮控福海往伊兄王煥章水井內傾倒煤油一案表，J181-019-42140，1924 年 10 月。)

21 *Yishi baihuabao*, June 2, 1917. (益世白話報 1917 年 6 月 2 日 第三版 京聞 洋井壟斷居奇。)

22 *Beijing baihuabao*, November 2, 1922. (北京白話報 1922 年 11 月 2 號 第三版 父子大打倒水夫。)

23 Beijing Municipal Archives, *Jingshi jingchating zhongyiqu guanyu tishu canyiyuan Tao Jingyu kong tiaoshuiren Zhang Xiuqing you touqie yifu xianyi yian*, J181-019-26862, September 12, 1920. (北京市檔案館：京市警察廳中一區關於提署參議員陶經武控挑水人張秀清有偷竊衣服嫌疑一案的呈，J181-019-26862，1920 年 9 月 12 日。)

24 "Beijing jingye diaocha," *Xinghua* 23, no. 32 (1926): 24–25. (北京井業調查，興華 1926, 23(32) ，第 24-25 頁。)

guild under the name of a carriers' association formed to represent their interests.[25] Liang's strategy turned out to be successful. In October, the Carriers' Association was finally approved and established its main office at Leshanli (樂善裏) in the Outer City, with Liang Chunhe serving as the head.[26]

The Carriers' Association limited its membership to natives of Shandong Province and set up compulsory rules and a supervisory committee. Members had to abide by the guild's rules, which set prices, wages, terms of credit, and so on. Any violators were subject to reprimands, fines, and even expulsion. In return, members would have backing for issues such as unjust lawsuits and disputes with customers or people in the same occupation. The Carriers' Association created eight branch offices in different parts of the city. Each wielded tremendous power through which the guild maintained rigid control over the water supply.[27] It enforced the spatial order for water carriers, each of whom held exclusive rights to supply water to a specific neighborhood. Even if the residents of a neighborhood wanted to switch their carrier, no external carriers would risk carrying water for them.[28] The guild solidified cohesion among water carriers throughout the city and, in the mid-1920s, incited them to protest against the expansion of piped water and to implement their own form of water governance.

2 The Politics of the Shoulder Pole

The Carriers' Association prepared the water carriers to reestablish their water regime. Water carriers began to launch coordinated operations by using shoulder poles as weapons to evict running water from their distribution territories and redefine their water regime in 1925. Water was used to advance group interests as modernization processes. In the meantime, the water carriers used the patriotism movement to legitimize their campaign against running water. The politics of the shoulder pole implied that water carriers directly employed organized violence to implement their own water governance.

---

25 *Shuntien Times*, April 21, 1925. (順天時報 1925 年 4 月 21 日 第七版 水夫運動結團體。)

26 *Shuntien Times*, October 14, 1925. (順天時報 1925 年 10 月 14 日 第七版 水夫仍躍躍欲試。)

27 *Chenbao*, June 10, 1926. (晨報 1926 年 6 月 10 日 第六版 水夫昨已經復工 官廳不準加價。)

28 *Shuntien Times*, April 29, 1924. (順天時報 1924 年 4 月 29 日第七版 水閥糞閥之大壟斷。)

At the end of 1924, to improve accessibility and expand its market, the Waterworks began sending its own non-guild carriers to deliver piped water.[29] This aggressive plan led to vehement opposition from the Shandong water carriers. On December 17, 1924, a Capital Waterworks carrier delivering water was ambushed by fifty well-water carriers from the guild. He escaped only because of the timely appearance of the police. Two days later, the Waterworks sent another two carriers, Zhang and Liu, to distribute water. Suddenly, fifty or sixty well-water carriers surrounded and beat them with shoulder poles. Liu tried to flee but was closely pursued on foot; he found shelter in one resident's house, but the water carriers broke in and began to beat him again. Thanks to the intervention of the homeowner, Liu suffered only minor injuries, but his co-worker Zhang suffered severe injuries, and all their working utensils such as their buckets and shoulder poles were smashed.[30]

Confronted with repeated violence and retaliation, Capital Waterworks had to dismiss its newly established water carriers and, in early 1925, it changed its strategy to install more internal pipes in order to expand its business. But this only intensified the conflicts between plumbers from the Waterworks and water carriers. To appease the water carriers, the Waterworks negotiated with thirty representatives of the well guild. In February 1925, after long and arduous bargaining, the two sides reached a provisional agreement in which water carriers would be allowed to sell forty thousand buckets of running water per day in the Outer City and the western part of the Inner City. In return, the Waterworks agreed not to install any more pipes. However, neither side signed the official document. Meanwhile, the Waterworks changed its regulations and increased water prices, which triggered a strike by the running-water carriers.

On April 1, 1925, in an effort to restrict the steady encroachment on water carriers' territories by the Waterworks, the well guild called a meeting at its guild hall and decided that all water carriers in the Outer City would strike, terminating their cooperation with Capital Waterworks.[31] Meanwhile, the guild called on water carriers to physically resist the Waterworks' efforts to install more pipes within their territories. This appeal brought an immediate response of support from carriers. On that same day, water carrier Le saw

29 "Chouban shuiche songshui youguan wenjian," in Beijingshi danganguan, ed., *Beijing zilaishui gongsi dangan shiliao* (Beijing: Beijing yanshan chubanshe, 1986,) 127–128. (籌辦水車送水有關文件，北京市檔案館等編，北京自來水公司檔案史料 北京：北京燕山出版社, 1986 年, 第 127–28 頁。)

30 Beijing Municipal Archives, *Beijing zilaishui gongsi zhi jingshi defang jiachating*, J067-1-222, December 20, 1924. (北京市檔案館：北京自來水公司致京師地方檢察廳，J067-1-222，1924 年 12 月 20 日。)

31 *Chenbao*, April 2, 1925. (晨報 1925 年 4 月 2 日 第六版 擔水夫將罷工。)

Capital Waterworks plumber Zhao installing internal pipes in resident Huang's house within Le's distribution territory. Le, together with more than one hundred carriers, stopped the installation.[32] The next day, carrier Fu succeeded in preventing a Waterworks plumber from installing pipe in resident Zhang's house within Fu's territory, and seized the Waterworks' pipes and tools.[33] In mid-April, water carrier Liu noticed running-water workers installing pipe in two households in his territory. He immediately assembled twenty water carriers to thwart the installation.[34] Another fight broke out when resident Liu, in water carrier Lin's distribution territory, applied for running water to be installed. As the trench to lay pipe was being excavated, one hundred water carriers carrying poles in their hands surrounded the plumbers; the project was brutally suspended.[35]

Backed by the well guild, the water carriers intensified their long-standing campaign against Capital Waterworks. On April 22, well-house owner Shao Hanqing, who was also an executive member of the well guild, led dozens of carriers in stopping three Waterworks employees from installing pipe in Nanluoguxiang (南鑼鼓巷). Two days later, counting on police protection, the same plumbers returned to resume their work in the trench. In a short while, more than twenty carriers with shovels in hand appeared, refilled the trench, and almost buried the workers.[36] In early May the frustrated Waterworks sued the Shandong water carriers at the police department on the grounds that they were undermining the running water business by impeding the installation of private pipes. The Waterworks also appealed to the police department to prohibit water carriers' illegal actions and provide protection to running-water

32 Beijing Municipal Archives, *Neiyou yiqu biaosong Le Jianren hangao zilaishui gongsi sixun Zhao Zishao duoshou shuiguandao yian*, J181-019-46843, April 1925. (北京市檔案館：內右一區表送樂建仁喊告自來水公司司巡趙子韶侵奪售水管道一案，J181-019-46843，1925 年 4 月。)

33 Beijing Municipal Archives, *Tiaoshuiren Fu Bingxiang hangao zilaishui gongsi sixun Zhao Zishao duoshou shuiye*, J181-19-46842, April 1925. (北京市檔案館：挑水人傅炳湘喊告自來水公司司巡趙子韶侵奪售水業，J181-19-46842，1925 年 4 月。)

34 Beijing Municipal Archives, *Chenjie chahuo tuishuifu Chen Yuling deng juzhong zunao zilaishui gongsi anshe shuiguan yian*, J181-19-47839, April 13, 1925. (北京市檔案館：呈解查獲推水夫陳玉齡等聚眾阻撓自來水公司安設水管一案， J181-19-47839，1925 年 4 月 13 日。)

35 Beijing Municipal Archives, *Waizuo sanqu biaosong Lin Yunjie zuzhi zilaishui anshe zhuanguan an*, J181-019-47838, April 15, 1925. (北京市檔案館：外左三區表送林允傑阻止自來水安設專管，J181-019-47838，1925 年 4 月 15 日。)

36 Beijing Municipal Archives, *Neizuo sanqu biaosong Shao Lijian zhushi Ma Qingfa zunao anzhuang zilaishuiguan an*, J181-019-47840, April 1925. (北京市檔案館：內左三區表送邵立儉主使馬清發等阻撓安裝自來水管案，J181-019-47840，1925 年 4 月。)

workers. Simultaneously, the water carriers also petitioned government ministries, the police, and the chambers of commerce for relief.[37]

Throughout May of 1925, fights between carriers and Waterworks plumbers in many of the city's streets slowed down the installation of piped water. On May 21, a running-water worker attempted to move and replace an old pipe in resident Yang's household. Water carrier Wang, along with numerous followers, swarmed into the yard and stopped the work. The next day, two carriers appeared just as running-water workers were getting ready to install pipe in a shop; half an hour later fifty or sixty carriers blocked the lane. Due to the intervention of the police, the water carriers did not beat the workers, but they did refill the trench and break the work cart.[38] These constant conflicts compelled the police to intervene, even though the police were themselves engaged in the water business through subleasing their wells to water carriers. On June 9, the Capital Police Department ordered water carriers not to obstruct private installations of running water. However, the well guild also petitioned the police department for the right to defend themselves, claiming that the tradition of water carriers had been rooted in the city of Beijing for hundreds of years. They had dug wells, built well houses, and supplied water to the city following long-established water distribution territories, and were thus an integral part of the city. The carriers claimed that the Waterworks had encroached on their territories and taken away their livelihoods. Under the pressure to make a living, they had been forced to compromise and sell water for the Waterworks, but the Waterworks had cheated them by breaking their contract and covertly installing pipe. The water carriers thus had no choice but to fight back, and the police seemed to accept the legitimacy of the water carriers' strategies to preserve themselves.[39]

The May Thirtieth Movement, a major labor and nationalist movement, presented the water carriers with an excellent opportunity to articulate their labor rights and legitimize their battle against running water. On May 30, 1925, the British Shanghai Municipal Police killed thirteen and wounded over fifty unarmed Chinese demonstrators in the International Settlement who were protesting the killing of a Chinese worker by a foreman at a Japanese cotton

37 *Shuntien Times*, May 4, 1925. (順天時報 1925 年 5 月 4 日 第七版 自來水公司控告山東水夫。)

38 Beijing Municipal Archives, *Jingshi zilaishui gongsi shang jingshi jingchating*, J067-1-222, May 22, 1925. (北京市檔案館：京師自來水公司上京師警察廳，J067-1-222，1925 年 5 月 22 日。)

39 Beijing Municipal Archives, *Beijing shuihang jingye tongren shang jingshi jingchating*, J067-1-222, June 1925. (北京市檔案館：北京水行井業同人上京師警察廳，J067-1-222, 1925 年 6 月。)

mill. The shootings sparked nationwide anti-imperialist demonstrations. The high tide of the May Thirtieth Movement in Beijing occurred in June, when students in Beijing mobilized laborers to protest. The members of the well guild responded enthusiastically. At the request of the student association, the guild mobilized water carriers to distribute pamphlets to residents' houses when delivering water.[40]

Water carriers used the pretext of nationalism by taking advantage of the May Thirtieth Movement to increase their violent opposition to Capital Waterworks and to validate their actions against running water. The water carriers were not necessarily concerned about national politics, but they were shrewd enough to understand the ways in which patriotism could be utilized for their local purposes and could be used to legitimate their violence. In June, water carriers suspended deliveries to English and Japanese clients in Beijing. On June 15, Beijing water carriers conducted a one-day strike in solidarity with Shanghai strikers, with four hundred water carriers from the southern and western sections of the city attending a citywide parade and holding a white flag displaying the name of the well guild in large characters.[41] The well guild even petitioned the Ministry of Foreign Affairs to declare war on Britain and Japan if negotiations were unfavorable to China, and claimed that four thousand water carriers in Beijing were willing to transport ammunition and to forage for the army to help it recapture Shanghai concessions and redeem China.[42]

As David Strand in *Rickshaw Beijing* notes, not all political actions in June 1925 were motivated by nationalistic sentiments. Some groups saw the freedom of assembly and action fostered by mass nationalism as a unique opportunity to press particular grievances. Nationalistic rhetoric and solidarity became a means for advancing group and class interests.[43] Thus the water carriers used the pretext of patriotism to legitimize the well guild and their campaign against running water: they exploited the May Thirtieth Movement to increase their influence and to limit the expansion of Capital Waterworks. After the strike on June 15, more water carriers joined in violent confrontations with the Waterworks. On June 17, water carrier Zhao stopped the pipe installation at

40 *Shenbao*, July 18, 1925. (申報 1925 年 7 月 18 日 第 7 版 京各界積極籌備國民大會。)

41 *Shuntien Times*, June 16, 1925. (順天時報 1925 年 6 月 16 日 第七版 北京五萬人遊行，水夫四百加入遊行。)

42 *Beijing shuijingye gonghui shuifu cheng*, 03-40-022-01, June 16, 1925, Archives at Institute of Modern History, Academia Sinica, Taipei. (中研院近史所檔案 ：03-40-022-01，北京水井業工會水夫呈，1925 年 6 月 16 日。)

43 David Strand, *Rickshaw Beijing: City People and Politics in the 1920s* (Berkeley: University of California Press, 1989), 188.

resident Zhang's house.[44] Water carrier Fu was sent by the well guild to prevent the Waterworks from installing pipe at another resident's house and was later arrested by the police. The well guild submitted bond to bail him out.[45]

The clash at Ping'anli (平安里) represents the peak of the water carriers' resistance during the May Thirtieth Movement. On June 21, the residents in the community of Ping'anli applied for a public hydrant by raising funds to purchase pipes. At their request, the Waterworks sent employees to install the hydrant. Water carrier Xie, who distributed water in this district, led forty carriers to prevent the installation.[46] The following day, the Waterworks resumed its work with police protection. Two hundred water carriers arrived to obstruct the project. The police called for backup and ordered the terrified workers to find shelter at the east gate of Ping'anli, but it was too late. The water carriers ignored the outnumbered police officers and assaulted the workers with shoulder poles, bricks, and ropes. They also took away the workers' pipes and tools. The panicked police were only able to arrest two carriers.[47] That same day, several hundred carriers from the Outer City, armed with shoulder poles, marched toward the offices of the Waterworks in the East City to protest pipe installations. Throughout June the water carriers continued to create havoc in the streets of the city. On June 24, a crowd of six hundred carriers, armed with poles and rods, clashed with Waterworks employees laying pipe in the West City, leading to fourteen injured workmen and a few arrests made by the police.[48] Four days later, water carriers severely injured a worker who was installing pipe in a private house in the eastern portion of the city.[49]

As the case of Ping'anli shows, the May Thirtieth Movement had given water carriers an opportunity to extend to the whole city conflicts with Capital

---

44 Beijing Municipal Archives, *Waizuo erqu biaosong shuifu Zhao Xipo shuaizhong zunao zilaishui gongsi anshe zhuanguan an*, J181-019-47841, June 1925. (北京市檔案館：外左二區表送送水夫趙喜坡率眾阻擾自來水公司安設專管案，J181-019-47841，1925 年 6 月。)

45 Beijing Municipal Archives, *Zilaishui gongsi gongtou Zhao Zishao konggao tiaoshuifu Fu Wenqing lanzu zuogong*, J181-19-47844, June 17, 1925. (北京市檔案館：自來水公司工頭趙子韶控告挑水夫傅文清攔阻作工案，J181-19-47844，1925 年 6 月 17 日。)

46 Beijing Municipal Archives, *Neiyou siqu biaosong Shuang Haiquan kong Xie Zhongsi zurao anshe zilaishuiguan an*, J181-019-47842, June 21, 1925. (北京市檔案館：內右四區表送雙海泉控解鐘思阻撓安設自來水管一案，J181-019-47842，1925 年 6 月 21 日。)

47 Beijing Municipal Archives, *Neiyou siqu biaosong Shen Lianyao zuzhong zurao zilaishui gongsi anshe shuiguan an*, J181-019-47843, June 22, 1925. (北京市檔案館：內右四區表送沈連堯等聚眾阻撓自來水公司安設水管等情一案，J181-019-47843, 1925 年 6 月 22 日。)

48 *North China Standard*, June 25, 1925, p. 8.

49 *Shuntien Times*, June 29, 1925. (順天時報 1925 年 6 月 29 日 第七版 水風潮大圖不休。)

Waterworks that were normally confined to neighborhoods. However, factionalism was an issue among the water carriers. The Waterworks had more intensive pipe networks in the eastern section of the city than in the western section, creating a spatial disparity among the water carriers. The water carriers in the piped communities of the eastern city who had been recruited by the Waterworks were more willing to collaborate with it; while water carriers in the western city, where running water was a new threat, were more committed to opposing the Waterworks. On June 18, the water carriers in the western portion of the Outer City assembled to participate in a scheduled march. While proceeding toward the government building designated as the focus of the protest, they suddenly veered north to attack well houses in the northeastern section of the Inner City and damaged four or five well houses belonging to fellow carriers with whom they had feuds.[50] The attacking carriers used the freedom to join public demonstrations as a cover to indulge in these indoor politics. The next day the victims of the water carriers asked other water carriers in the northeastern portion of the city to publicize this provocation. The irritated water carriers planned to retaliate but were stopped by cooler-headed water carriers who proposed investigating the ringleader and then took steps to do so.[51] On June 22, several hundred water carriers holding shoulder poles marched toward the offices of the Waterworks in the Inner City. As they paraded, they coerced carriers in the northeast of the city who had been their adversaries to join the march.[52]

While participation in the May Thirtieth Movement seemed to normalize the carriers' violence on the pretext of nationalism and patriotism, they also used the mismanagement of the Waterworks' sanitation as an excuse to resist running water. Supplying water directly from the river had raised problems of cleanliness. As early as June 1922, residents had already discovered that running water had taken on a yellowish color and its taste was exceedingly unappealing. If used to make tea, the tea soon turned black.[53] The newspaper speculated that the mains, long neglected and in disrepair, had a great deal to do with the impurity of the water.[54] This accumulated negligence finally led to a crisis about the purity of running water in 1925. In June, the users of running water found that their tap water was turbid, yellow, and even contained

50 *Shuntien Times*, June 23, 1925. (順天時報 1925 年 6 月 23 日 第七版 水夫亦假愛國運動美名，實際與自來水公司爭利。)

51 *Shuntien Times*, June 20, 1925. (順天時報 1925 年 6 月 20 日 第七版 京城水夫之黨派潮。)

52 *Shuntien Times*, June 23, 1925. (順天時報 1925 年 6 月 23 日 第七版。)

53 *Shuntien Times*, June 29, 1922. (順天時報 1922 年 6 月 29 日 第七版 自來水之質危險。)

54 *Shuntien Times*, July 13, 1922. (順天時報 1922 年 7 月 13 日 第七版 自來水宜加修理。)

sand and feathers. The newspaper reported that the settling ponds of the Waterworks, which were supposed to treat and purify the water, were useless. In addition, the sand filtering-beds, which were an old type that worked very slowly, had not been maintained and were therefore more or less inoperative. The water was so contaminated that one doctor discovered that all his diarrhea patients were the consumers of running water. People finally asked the state to intervene.[55] At the residents' request, the Central Plague Prevention Bureau tested the water that came through the pipes of the Waterworks and detected 600 microbes in every fifteen drops of water because the Waterworks was lax in cleaning its filter beds. The water was not any cleaner than the water of the modern wells, although a high charge was levied for its supply. As a result of this astonishing finding, the Bureau concluded that the water supplied by the Waterworks was extremely dangerous to drink.[56] The police department ordered the Waterworks to improve its sanitation facilities, and residents suspended their use of running water and switched to well water. The well guild called a meeting to discuss methods for using the crisis to solicit clients.[57]

After the June conflict in Ping'anli, the police convened a mediation session and proposed a compromise in which the Waterworks was given the right to lay pipe to shops and homes that used thirty tubs or more of water per day, while the water carriers were given a claim to those who used less. The Waterworks suspended installations to users of less than thirty tubs per day, but the water carriers were still dissatisfied and continued to clash with Waterworks plumbers. In September, thirteen running-water workers were digging a trench to lay pipe in front of a private home. An executive member of the well guild, Yao Hanqing, led sixty carriers to swarm the street and fight with the workers. The outnumbered workers fled and the carriers carried off all the pipes and smashed the work carts prior to the arrival of the police.[58] In October, the same Yao Hanqing instructed carrier Guo to lead dozens of water carriers with shoulder poles, iron rods, and ropes to prevent workers from installing pipe in a resident's house.[59] In December, the Waterworks planned to install private pipe in carrier Yue's territory; Yue asked for thirty *yuan* as compensation for

---

55 *Chenbao*, June 30, 1925. (晨報 1925 年 6 月 30 日 第六版 自來水與瀉病。)

56 "Peking Waterworks," *The North-China Herald*, August 15, 1925.

57 *Chenbao*, August 9, 1925. (晨報 1925 年 8 月日 第七版 水屋議減水價 招攬自來水用戶。)

58 *Chenbao*, September 30, 1925. (晨報 1925 年 9 月 30 日 第七版 水夫聚毆自來水工人。)

59 Beijing Municipal Archives, *Tiaoshuifu Guo Xizeng juzhong zunao zilaishui gongsi gongjiang anshe shuiguan an*, J181-19-47847, October 15, 1925. (北京市檔案館：挑水夫郭喜增聚眾阻撓自來水公司工匠安設水管一案，J181-19-47847，1925 年 10 月 15 日。)

each piped household. The Waterworks only agreed to one *yuan*, and thus Yue resisted any pipe installation within his distribution territory.[60]

In spring 1926, the Waterworks slowed down its pipe installation because of a water famine. The pumps for running water were fueled by coal, which was transported by rail; but military conflicts among the warlords on the outskirts of Beijing had cut off the railway and suspended the coal supply. Throughout March the Waterworks constantly published notices in the newspaper to remind customers to conserve water.[61] The adversity because of the lack of running water brought about a transient truce between the water carriers and the Waterworks. However, the violence again escalated following the increased consumption of water in the summer. At the end of May, when a running-water worker was laying pipe, an executive member of the well guild, Yao Hanqing, accompanied by dozens of water carriers, compelled the workers to refill the trench.[62] On June 1, well-house owner Wang Yuqian, after consulting with the Carriers' Association, gathered seven well-house owners and fifty water carriers to prevent the installation of pipes.[63]

The clash in Jingzhongmiao (精忠廟) signified the second peak of the water carriers' confrontations with the Waterworks. On the morning of June 7, 1926, water carrier Zheng saw Waterworks workers installing pipe at two glass stores within his distribution territory in Jingzhongmiao. He deliberately proclaimed the shortcomings of running water on the spot, which irritated the plumber Huang. Huang argued with Zheng and was beaten with a shoulder pole. Later, two more water carriers joined the fight and smashed the workers' utensils.[64] The police arrived and detained three water carriers. With his carriers

60 Beijing Municipal Archives, *Jingye tiaoshuifu Yue Dianwen deng zuzhi zilaishui gongsi gongren anshe Zhang Yuantong zhuzhai shuiguan yian*, J181-19-47848, December 12, 1925. (北京市檔案館：井業挑水夫嶽殿文等阻止自來水公司工人安設張元同住宅水管一案，J181-19-47848，1925 年 12 月 12 日。)

61 *Shuntien Times*, March 6–April 1, 1926. (順天時報 1926 年 3 月 6日–4月 1 日 第一版 北京自來水有限公司緊要通知。)

62 Beijing Municipal Archives, *Zilaishui gongtou Ning Qianyi kong daoshuifu Yao Hanqing deng zuzhong zunao anshe shuiguan yian*, J181-19-50989, June 1926. (北京市檔案館：自來水工頭寧潛儀控倒水夫姚漢卿等聚眾阻撓安設水管一案，J181-19-50989，1926 年 6 月。)

63 Beijing Municipal Archives, *Waizuo yiqu biaosong shuijing zhanggui Wang Yuqian shuaizhong zunao zilaishui gongsi anzhuang shuiguan yian*, J181-019-52012, June 1926. (北京市檔案館：外左一區表送水井掌櫃王雨乾糾眾阻擾自來水公司安裝水管一案，J181-019-52012，1926 年 6 月。)

64 Beijing Municipal Archives, *Waizuo wuqu chengsong Zhou Hengyu deng zuzhi Ning Qianyi anzhuang zilaishuiguan yian*, J181-019-50990, June 1926. (北京市檔案館：外左五區呈送周恒玉等阻止寧潛儀安裝自來水管等情一案，J181-019-50990，1926 年 6 月。)

in custody, well-house owner Yang appealed to the Carriers' Association for a strike and was arrested as well. In the evening, the Association tried to bail them out but was turned down by the police.[65] The next day, the head of the Carriers' Association, Liang, called a meeting with 103 well-house owners to discuss resisting the Waterworks and increasing water prices. As justification for its actions, the Carriers' Association presented three current drawbacks of running water. First, running water drawn from the Sun River without purification was turbid. Moreover, the recent military conflicts had left dead bodies decomposing in the river, which dramatically contributed to its pollution. Second, the Waterworks was experiencing a shortage of the coal that was used to fire its steam pump, and this thwarted its daily business. If residents relied exclusively on running water, they would immediately experience stoppages. Finally, the price of running water for two buckets was three coppers without delivery, while well water was the same price and included delivery.

The June clash became a suitable occasion for the ventilation of water carriers' general grievances; the Carriers' Association decided to launch a strike and petitioned the police department and municipal office. They demanded that the city's water supply should henceforth be solely furnished by the carriers. They also stated that when supplies were resumed, prices would be raised. The police department denied this petition and ordered the guild to immediately resume supplying water. In the evening, the water carriers had another meeting in which the majority insisted on striking.[66] On June 9, the water carriers held a one-day strike and decided to increase the water price for one load of two buckets from two coppers to five, but this action was revoked by the police department.[67] Yet, the failure of the movement by no means damped the hostility of the carriers. Nor did an increase in rates, by one copper per brace of buckets, end their agitation.[68] When running-water plumbers installed pipe at a grocery store in July, well-house owner Wang Yuqian once again led dozens of carriers to attack the workers and take their wrenches.[69]

---

65 Beijing Municipal Archives, *Waiyou siqu chengsong songshuifu Yang Zijiu shanhuo tongye bagong yian*, J181-017-03034, June 1926. (北京市檔案館：外右四區呈送送水夫楊子久煽惑同業罷工一案，J181-017-03034，1926 年 6 月。)

66 *Chenbao*, June 9, 1926. (晨報 1926 年 6 月 9 日 第六版 昨日全城水井業罷工。)

67 *Chenbao*, June 10, 1926. (晨報 1926 年 6 月 10 日 第六版 水夫昨已經復工 官廳不準加價。)

68 "Water Supply: The Old and the New in China," *South China Morning Post*, July 29, 1926, p. 6.

69 Beijing Municipal Archives, *Chengjie daoshuifu Gao Jiancai zunao zilaishui gongsi anzhuang shuiguan bing oushang gongren deng*, J181-19-51237, July 26, 1926. (北京市檔案館：呈解倒水夫高建才阻撓自來水公司安裝水管並毆傷工人等，J181-19-51237，1926 年 7 月 26 日。)

The years 1925 and 1926 witnessed the solidarity of the water carriers, who employed various strategies to implement their water governance by thwarting the expansion of piped water and keeping firm control over 80 percent of Beijing's water supply. The water carriers expelled the running-water workers through violence and stopped distributing water to households that had requested pipe installation. This was done so frequently that the police had to resort to mediation rather than simple repression. When the police intervened, the Carriers' Association would assign one carrier to plead guilty if charged in a lawsuit. These guerrilla legal actions swamped the courts with water cases and also kept the Waterworks exhausted from coping with suits, which proved a very effective approach to manipulating the law in order to resist the Waterworks. Moreover, because of the inconsistency of running water, the police department was unable to apply extreme pressure on the water carriers. For example, in December 1925, water carrier Yue impeded pipe installation and sabotaged tools and pipe in a resident's house. According to the police, although this was illegal, the Waterworks had not provided a solution to the water carriers' loss of livelihood, which was also improper.[70] In general, the government wielded a light hand by mediating with water carriers instead of punishing them, which also contributed to the water carriers' increasing power.[71]

3 The State and the Water Lords

Regarding the state's water governance in the Beiyang period, owing to the unstable and tumultuous warlord state, Beijing's underfunded municipal government was neither capable of nor willing to take responsibility for providing clean water to the populace, so it positioned clean water as a private commodity rather than a public good. The government consistently saw its responsibility only as mediating between the water carriers, Capital Waterworks, and city dwellers; and this laissez-faire approach gave rise to the transformation of the water carriers into water lords.

To tackle the monopoly of the water carriers, the state realized that it had to address the water distribution territories, which were the embodiment of

70 Beijing Municipal Archives, *Jingye tiaoshuifu Yue Dianwen deng zuzhi zilaishui gongsi gongren anshe Zhang Yuantong zhuzhai, zaihui shuiguan ji jiaqu an*, J181-19-47848, December 12, 1925. (北京市檔案館：井業挑水夫嶽殿文等阻止自來水公司工人安設張元同住宅水管，砸毀水管及家具等一案，J181-19-47848，1925 年 12 月 12 日。)

71 *Shuntien Times*, June 10, 1926. (順天時報 1926 年 6 月 10 日 第七版 水夫蠻打自來水工人。)

the carriers' exclusive rights. After the 1911 revolution, the new Republican government had proposed getting rid of water distribution territories to make way for modern piped water. In a lawsuit over these territories, the judge had questioned their legitimacy and suggested that the government retake them to make running water prosperous.[72] However, this was wishful thinking by the court. The state deferred to the water carriers for two reasons. First, the water carriers played an essential role in maintaining the water supply, which the unreliable system for running water was unable to replace. Second, before outlawing the territories, the state had to consider the livelihood of thousands of water carriers.

Throughout the 1910s, the state, faced with warlord feuds, had very limited resources for dealing with the intricate politics of the water distribution territories. Instead, it did nothing more than issue some general regulations for wells. In 1914, the police department established a hygiene office to control and oversee public and private wells.[73] Three years later, the municipal office evaluated, classified, and ranked all the wells in the city according to quality.[74] In 1923, in a piece of property legislation, the Supreme Court revoked the right to own water distribution territories, outlawed water carriers' exclusive right to supply water to a specific territory, and endowed residents with freedom of choice in drinking water. However, there is no evidence to indicate that this legislation was successfully implemented or that water carriers complied with it.

The state failed to push back, which resulted in the power and violence of the water carriers as water lords after 1923 when the Waterworks shattered its cooperation with the carriers. In one petition based on this legislation, clock shop owner Liu represented twenty stores in Andingmen Street (安定門內大街) in an appeal to the police department to punish well-house owner Du, who had increased water prices and monopolized the water supply. However, the police department concluded there was no evidence that Du interfered with residents' freedom of choice in drinking water; rather, Liu had played the public opinion card by fabricating a petition to accuse Du.[75]

---

72 Beijing Municipal Archives, *Wei Yuting su Wang Yongqing shuidao jiue yian*, J181-19-2010, April 1913. (北京市檔案館：魏玉亭訴王永清水道糾葛一案，J181-19-2010，1913 年 4 月。)

73 *Shenbao*, September 8, 1914. (申報 1914 年 9 月 8 日 第 10 版 京師警察分科職掌規則（續）。)

74 Jingdu shizheng gongsuo, ed. *Jingdu shizheng huilan* (Beijing: Jinghua shuju, 1919), 224–226. (京都市政公所編：《京都市政匯覽》，北京：京華書局，1919 年，224–226 。)

75 Beijing Municipal Archives, *Du Xiuchun bachi shuidao an*, J181-18-15643, July 1923. (北京市檔案館：杜秀春把持水道案，J181-18-15643，1923 年 7 月。)

Confronted with the violence of water carriers in the 1920s, the government's view was that it had only limited responsibility for adjudicating disputes between the Waterworks and the carriers. In other words, it essentially tolerated the water carriers' violence. More importantly, the municipal government itself was engaged in business with the water carriers by subleasing governmental wells to them. Since the Waterworks never pressurized their pipes at night for reasons of economy, the fire brigades were unable to draw water from dry hydrants when fires broke out at night. So the police department drilled at least 480 wells in its twenty districts for the purposes of firefighting and sanitation. To raise funds, it subleased a fraction of these wells to water carriers who sold the water.[76] One government well west of Caishikou, drilled and owned by the police department, was subleased to a water carrier who added cisterns and sewers to sell the water.[77] Accordingly, the police department had no intention of expelling the water carriers.

Even as water carriers repeatedly impeded pipe installations for running water in 1925, the police department strove to find a compromise between the carriers and the Waterworks. In April, water carrier Lin, together with one hundred carriers, interrupted a pipe installation. The court concluded that running water had impacted Lin's business, so his actions were justified. Lin was released after being admonished.[78] In June, concerned with the increasing violence of the water carriers, the police department issued an order prohibiting the carriers' radical campaign against running water. In the meanwhile, the police made a request to the Waterworks that it cancel its plan for pipe installations.[79]

During the warlord period, with the help of Peking Union Medical College under the patronage of the Rockefeller Foundation, the municipal government finally began to scrutinize drinking water as one of the main concerns of hygiene. In 1925, the Department of Hygiene and Public Health, supervised by John B. Grant (1890–1962), was established using the model of Johns Hopkins

76 "Beijing jingye diaocha," *Xinghua* 23, no. 32 (1926): 24. (北京井業調查，興華 1926，23 (32)，第 24 頁。)

77 *Shuntien Times*, December 28, 1927. (順天時報 1927 年 12 月 28 日第七版 承租官井售水之糾葛。)

78 Beijing Municipal Archives, *Waizuo sanqu biaosong Lin Yunjie zuzhi zilaishui anshe zhuanguan an*, J181-019-47838, April 15, 1925. (北京市檔案館：外左三區表送林允傑阻止自來水安設專管案，J181-019-47838，1925 年 4 月 15 日。)

79 Beijing Municipal Archives, *Jingshi jingchating ling*, J067-1-222, June 1925. (北京市檔案館：京師警察廳令，J067-1-222，1925 年 6 月。)

University.[80] But little was done during the period of the civil wars until Feng Yuxiang (馮玉祥, 1882–1948), the Christian general and warlord, captured the Beijing region in 1924 and appointed a follower of his, a lawyer by the name of Xue Dubi (薛篤弼, 1892–1973), as governor of the metropolitan area. The Hygiene Department under John Grant had no difficulty in establishing friendly relations with Xue, and the department consequently completely reorganized its relationship with the police and other authorities.[81] The connections John Grant nurtured with Beijing's municipal government helped his Hygiene Department establish a public health station that serviced a population of 100,000 in the eastern part of Beijing.

Under this influence, Beijing's municipal government established three additional public health demonstration areas. In September 1925, the Central Plague Prevention Bureau requested that the Beijing police department survey the health habits of inhabitants, including their drinking water.[82] Three police officers worked for a month, investigating one house at a time in a portion of a Beijing ward. They found polluted drinking water and poor water-closet sanitation, which together led to the excessive breeding of flies.[83] In August 1925, the police department urged the Waterworks to improve its sanitation and placed rigid regulations on the use of street standpipes because of the high concentration of microbes in running water.[84] From that time forward, during the summer, the examination of running water became a regular practice of the police department.[85] The police department also admonished well owners that all public and private wells should add well curbs and covers to prevent

80 John Grant was born to Canadian missionaries in China and earned his medical degree from the University of Michigan and a master's degree in public health from Johns Hopkins University. He was assigned to China by the Rockefeller Foundation, where he became a professor and later the head of hygiene and public health at Peking Union Medical College. Mary Brown Bullock, *An American Transplant: The Rockefeller Foundation and Peking Union Medical College* (Berkeley: University of California Press, 1980).

81 "The Reminiscences of Doctor John B. Grant," *New York Times Oral History Program,* Columbia University Oral History Collection, Part 3, no. 195, Medicine and Public Health (Glen Rock, NJ: Microfilm Corporation of America, 1976), 182.

82 Chinese Second Historical Archives, *Beijing Nanjing liangdi juban pumie cangying shixiang youguan wenjian*, 1036–80, May 15, 1925. (中國第二歷史檔案館藏中央防疫處檔案，1036–80，北京南京兩地舉辦撲滅蒼蠅事項有關文件, 1925 年 5 月 15 日, 中央防疫處致外右一區警察署。)

83 CMB, Inc., box 66, folder 465 (Health Station, 1924–1926), Rockefeller Archive Center, New York.

84 *Shuntien Times*, August 9, 1925. (順天時報 1925 年 8 月 9 日京師警廳取締自來水。)

85 *Shijie ribao*, October 5, 1926. (世界日報 1926 年 10 月 5 日 第六版 公共衛生事務所建議改良自來水。)

dust, and that owners should keep wells and flumes clean.[86] With their help, the Central Plague Prevention Bureau investigated the wells and running water in Beijing in 1927.[87]

The hygienic movement apparently was unable to restrain the growing power of water regime. In the summer of 1927, with the increasing hot weather, water carriers continued to raise the price of water.[88] The police department reiterated the order to outlaw any increases or suspension of the water supply, and violators were subject to severe punishment.[89] As one editorial complained, water carriers were one of the problems suffered by Beijing residents. But the government disregarded and underrated the seriousness of the problem.[90]

4 The Residents and the Water Lords

Despite the presence of running water, it was very unreliable; ongoing problems with the pumping apparatus, clogged intakes, and other malfunctions meant that residents still needed water carriers. At the outset, Capital Waterworks was under the patronage of Yuan Shikai. After Yuan's death in 1916, the Waterworks went steadily downhill as warlords and other prominent actors impeded its development. When the warlord Zhang Zuolin (張作霖, 1875–1928) was in control of Beijing from 1926 to 1928, he never paid his running water bill, which was 800 *yuan* per month. Important figures such as Yang Yuting (楊宇霆, 1886–1929), a Fengtien clique warlord, and Wellington Koo (顧維鈞, 1888–1985), an influential diplomat and the Minister of Foreign Affairs, did not pay the Waterworks a single cent over three years. Because of widespread nonpayment, the Waterworks had 100,000 accounts in arrears in 1928 alone. Moreover, the frequent coal shortages arising from suspension of railway services by warlords' military conflicts led to the scarcity of running

86 "Jingshi jingchating gaojie zhuyi shuijing weisheng," *Shizheng jikan*, no. 1 (October 1925), 27. (市政季刊 1925 年 10月 第 1 期 京師警察廳告誡註意水井清理，第 27 頁。)

87 Chinese Second Historical Archives, *Jingshi gequ shuijing koushu yilan ji jiancha jingshui zilaishui shixiang youguan wenshu*, 1036–81, July 13, 1926. (中國第二歷史檔案館藏中央防疫處檔案 1036–81，1926 年 7 月 13 日中央防疫處致內務部，京師各區水井口數一覽及檢查井水自來水事項有關文書，內務部致中央防疫處。)

88 *Shuntien Times*, April 25, 1927. (順天時報 1927 年 4 月 25 日 第七版 北京內外城挑水夫亦齊行漲價。)

89 *Shijie ribao*, October 16, 1927. (世界日報 1927 年 10 月 16 日 第七版 警察廳取締挑水夫 不準任意增價。)

90 *Shuntien Times*, November 19, 1927. (順天時報 1927 年 11 月 19 日 第七版 社會鏡。)

water.[91] Newspaper reports of running-water famines were very common in the warlord period of Beijing.

The inhabitants had no recourse but to turn to water carriers and well water when running water was unavailable. But residents could not choose their water carrier within a water distribution territory. After the Qing period, external water carriers were not allowed to distribute in others' territories without the permission of the water carrier who controlled the territory; this was considered unofficial law within the water regime. This exclusive right to the water supply caused tensions between residents and water carriers. For instance, resident Yang tried to fire his water carrier Wei and invite another water carrier to deliver his water. However, when the new carrier was carrying water to Yang's house, Wei seized the new carrier's water wheelbarrow.[92] Another example involved a new resident who did not know the rules. When he saw water carrier Liu and his water cart pass by, he tried to buy water from him. Unfortunately, his own carrier saw this, which led to a conflict.[93]

During the mid-1920s, the water carriers not only launched a bitter campaign against running water, they also strove to transfer to the residents the pressure they felt from competing with running water. The water carriers consistently employed the method of increasing water rates, especially in the summer; as a result, the price of water rose abnormally high. Formerly, the price per pair of buckets was one and a half coppers and weighed 30 kilos. In addition, the consumers had to pay extra tips on festival days. By 1926, carriers were charging three coppers for one load of two buckets. Moreover, carriers often milked the city's newcomers to the tune of 6 to 10 coppers for each load; if they did not pay, no water was supplied. Carriers were supported in their intimidating behavior by their compatriots, who refused to undertake a contract when a dispute arose between supplier and consumer; the residents were therefore coerced into submitting to any exorbitant demand.[94]

The motivation of water carriers for increasing water rates was not only to make more profit from selling water, but also to increase the value of each water distribution territory. For example, one water carrier purchased a territory at thirty *yuan*; after he increased water prices in the summer he could sell the territory for a price of fifty *yuan*. The new owner who took over would

91 *Jingbao*, August 22, 1928. (京報 1928 年 8 月 22 日 第六版 自來水公司收入銳減。)

92 Beijing Municipal Archives, *Neiyou sanqu biaosong An Zitian deng kong Wei Dianchen qinzhan shuiche yian*, J181-019-21909, February 9, 1918. (北京市檔案館：內右三區表送安子田等控魏殿臣侵占水車一案，J181-019-21909，1918 年 2 月 9 日。)

93 *Shijie ribao*, December 11, 1927. (世界日報 1927 年 12 月 11 日 第七版 水夫爭水道。)

94 "Water Supply: The Old and the New in China," *South China Morning Post*, July 29, 1926, p. 6.

continue the same practice, causing water prices to increase year by year.[95] In April 1925, water carriers in the western portion of the city even increased the price of well water to four coppers for two buckets. The residents considered the price too high and tried to negotiate with the water carriers, but the water carriers rejected their appeal and even threatened to suspend their water supply, which greatly distressed the lower-class residents.[96] Meanwhile the water carriers utilized another trick, which was to shrink the volume of their buckets by one-third. In the 1920s the price of two buckets of water was doubled, while the weight of two buckets of water decreased from thirty to twenty or even fifteen kilos.[97]

Within a water distribution territory, the suspension of the water supply was an important tool of manipulation used by the water carriers to coerce residents. Since drinking water was a daily necessity, the residents would soon submit if their water was cut off. Resident Li complained about the increasing price to her water carrier Ren. Ren then refused to distribute water to her. Li went to argue with him at the well house, but Ren threatened to beat her with his shoulder pole.[98] On top of increasing prices and suspending the water supply, violence was a weapon commonly used by water carriers to deal with residents. The militancy of the water carriers was so deeply rooted in residents' minds that it was included in Beijing's folklore: It was said that there was an ancient statue of a Chinese unicorn (*kylin*) at Tieshizi Hutong (鐵獅子胡同). It became a spirit and often stole drinking water from nearby wells at night. Water carriers wounded its head with a carrying pole.[99] As another example of water carrier violence, resident Wu was a military man and his water carrier failed to distribute water one day because of heavy rain. Furious, Wu went to the well house to argue with the water carriers, but was beaten by several of them.[100] If even a military man could fall victim to water carriers' violence, how much more susceptible would the ordinary inhabitant be? Such violence was possible because of the water carriers' solidarity enabled by the Carriers' Association. For example, water carrier Zhou had an altercation with one

95 *Shuntien Times*, December 21, 1927. (順天時報 1927 年 12 月 21 日 第七版 水閥仍積極保持權利。)

96 *Shuntien Times*, April 5, 1925. (順天時報 1925 年 4 月 5 日 第七版 水夫實行漲水價。)

97 *Shuntien Times*, July 8, 1927. (順天時報 1927 年 7 月 8 日 第七版 北京水夫運用漲價之手段 縮減水量以謀漁利。)

98 *Shijie ribao*, October 8, 1927. (世界日報 1927 年 10 月 8 日 第七版 挑水夫用武 要打用戶。)

99 *Aiguo baihuabao*, July 1, 1920. (愛國白話報 1920 年 7 月 1 日 第四版 謠言不值一笑。)

100 *Chenbao*, July 27, 1925. (晨報 1925 年 7 月 27 日 第七版 軍人水夫雨中大戰 爲缺水。)

customer over the price of water that drew hundreds of carriers from the western part of the city to the scene.[101]

Residents usually resorted to four methods when grappling with water lords, with varying degrees of success. The first was to install of running water. For example, in April 1925 water carriers increased water rates on the pretext of a hot summer. As a result, the residents proposed amassing funds and installing a public faucet in the neighborhood to avoid the constant bullying of the insatiable water carriers.[102] Nevertheless, the water carriers sought to impede the installation of piped water, especially, as we have seen, after the May Thirtieth Movement in 1925. For example, resident Han applied for the installation of piped water and made a down payment, but the installation was interrupted by water carriers.[103] Even powerful individual residents were unable to counter the water carriers. Zhang Yuntong, who was the chief judge of a superior court, intended to install running water pipes in his house. The Waterworks agreed to dispatch workers after Zhang secured police protection, but numerous water carriers appeared and interrupted the project. However, Judge Zhang was unwilling to give up and even invited two soldiers to safeguard his yard. With Zhang's promise, the reluctant Waterworks again sent its workers to resume installation. As soon as they began to work, twenty water carriers broke into the yard and smashed all the tools and pipes. Judge Zhang had to pay the 200-yuan loss to the Waterworks out of his own pocket. It seems that the installation of running water, as the first method to curtail water carriers' power, was generally unsuccessful.[104]

The second way to deal with water carriers was through the power of the neighborhoods. The gentry class and merchants, who possessed more social resources and had influential connections with police and other authorities, played an important role in negotiations as the patrons of the community. One interesting piece of evidence comes from Jiaochang liutiao (教場六條) where a foreign well was situated, but this well was adjacent to an old cemetery and dung field. Moreover, the borehole of the well had collapsed and the water was

101 *Shuntien Times*, June 23, 1925. (順天時報 1925 年 6 月 23 日 第七版。)

102 *Shuntien Times*, April 21, 1925. (順天時報 1925 年 4 月 21 日 第七版 水夫運動結團體。)

103 Beijing Municipal Archives, *Waiyou wuqu biaosong Yao Hanqing juzhong qiangzu zilaishuigong gongsi anzhuan shuiguan yian*, J181-019-47846, September 1925. (北京市檔案館：外右五區表送姚翰清聚眾強阻自來水公司安裝水管一案，J181-019-47846, 1925 年 9 月。)

104 Beijing Municipal Archives, *Jingye tiaoshuifu Yue Dianwen deng zuzhi zilaishui gongsi gongren anshe Zhang Yuantong zhuzhai shuiguan yian*, J181-19-47848, December 12, 1925. (北京市檔案館：井業挑水夫嶽殿文等阻止自來水公司工人安設張元同住宅水管一案，J181-19-47848，1925 年 12 月 12 日。)

extremely filthy. In the summer the inhabitants who drank this water suffered from dizziness and diarrhea; some even died from drinking it. Chen Ziheng, a member of the gentry, investigated and found that the ruined well collected rainwater from the adjoining cemetery and dung yard; so Chen gathered the neighbors to press the owner to repair it, but the owner refused. Chen then turned to the police department and this led to success.[105] However, because running water was a symbol of social status, it often divided society when elite families acquired it. The absence of elites dramatically weakened the ability of the neighborhoods to counterbalance a monopoly by water carriers. In such cases, the residents would resort to direct collective action to restrain the power of water carriers. Water carrier Li sold water in Anyuan Hutong (按院胡同), but he was disliked in the neighborhood. One day, the residents noticed that Li's water was odorous, so they forced Li to clean the well and they found a decayed dog. To appease the offended customers, the blameworthy Li apologized for his gross negligence and promised to buy a well cover and keep a close eye on the well in the future.[106]

The third method used by residents was petitioning the authorities to restrain the water carriers' power. In the 1920s, this power increased, and the residents continued to appeal to the authorities to overturn the water carriers' monopoly. One example is the clock-shop owner Liu, representing twenty stores and shops in Andingmen Street, who appealed to the police department. The shop owners accused water carrier Du of monopolizing the water supply and increasing prices.[107] In 1926, water carriers took advantage of the summer heat to increase water prices. In the meantime, they decreased the volume of their buckets. The poor households tried to argue with them, but the rude and indefensible carriers were indifferent to the residents' attempts to negotiate and, as a threat, immediately suspended the water supply. In response, the residents continued petitioning the police department to outlaw water carriers, but to no avail.[108]

The fourth approach for coping with water lords was well drilling. To avoid water carriers' bullying, more residents turned to drilling their own wells. But, for many individuals, it was not economically feasible to do this. This was especially true for the large number of tenant families who had no intention

105 *Chenbao*, August 31, 1924. (晨報 1924 年 8 月 31 日 第七版 井水不潔有礙衛生。)

106 *Chenbao*, August 25, 1924. (晨報 1924 年 8 月 25 日 第七版 死狗長眠水井中。)

107 Beijing Municipal Archives, *Du Xiuchun bachi shuidao an*, J181-18-15643, July 1923. (北京市檔案館：杜秀春把持水道案，J181-18-15643，1923 年 7 月。)

108 *Shuntien Times*, June 17, 1926. (順天時報 1926 年 6 月 17 日 第七版 警廳查禁 怪水夫因增漲水價。)

of opening wells in their landlords' yards. Moreover, the high rate of urban poverty thwarted the residents' impulse to drill wells. Only affluent families and institutions could afford to drill their own wells. For example, Yenching University, a preeminent Christian University in Beijing, had a well sunk to a depth of 120 feet by its geology professor George B. Barbour (1890–1977) in 1924, who supervised well drillings to relieve the drought in northern China under the patronage of the China International Famine Relief Commission in 1921.[109] This practice, combined with professional geological theory, made Barbour an expert in well-drilling.[110] On the whole, the limited number of private wells was not enough to dilute the water carriers' power.

5 Conclusion

The water regime in the period of 1924 to 1927 presents a complex picture since many actors—including water carriers, the Waterworks, the municipal government, and the residents—engaged with and were at odds with each other over the distribution of water. Cooperation between water carriers and the Waterworks broke down in 1924 when the Waterworks began to install internal pipes in households. In response, water carriers reorganized their well guild to solidify their power and articulate their interests through the manipulation of political movements, especially the May Thirtieth Movement of 1925. Their participation in national politics legitimized the carriers' violence against the expansion of piped water and helped them keep firm control over Beijing's water supply under the politics of shoulder poles. In this way, the water carriers built the citywide regime to implement their water governance.

---

109 George B. Barbour, "China International Famine Relief Commission to Barbour," 5/12/1921, in *In China When…* (Cincinnati, OH: University Publications, University of Cincinnati, 1975), 27–30, 62–63.

110 Based on these well-drilling experiences, Barbour published a few journal articles on the water supply in the Beijing area. George B. Barbour, "Water Supply of Tingchow and Shuntehfu Area," *Journal of the Association of Chinese and American Engineers* 3, no. 9 (October, 1922): 8ff; George B. Barbour, "Artesian Well Supply at Haitieu," *Journal of the Association of Chinese and American Engineers* 2, no. 1 (1924): 18ff; George B. Barbour, "Deep Wells in the Peking Area," *Bulletin of the Geological Society of China* 3, no. 2 (1924): 127–38; George B. Barbour, "Further Data Regarding Deep Wells in the Peking Area," *Bulletin of the Geological Society of China* 9, no. 1 (1930): 49–58. In addition to his own research, Barbour supervised his students in investigating Beijing's wells and water, see Chou Tung Pi 周同璧, "The Bitter Water and Sweet Water in Peking," Graduation Paper, Yenching University, 1926, Special Collection, Peking University Library.

The municipal government never succeeded in controlling the water carriers; rather, the government saw its responsibility in water governance as meditating between the water carriers and the Waterworks. Instead of municipalizing water and making it a public good, the government treated it as a private commodity. Moreover, the municipal government was intimately involved in the water carriers' business through its subleasing of government wells to them. To a large extent, its laissez-faire approach to the city's water provision and lack of interest in expelling the indispensable water carriers contributed to the making of the water lords. In addition, access to piped water divided society by extracting elite clients. This weakened the collective power of the neighborhoods, which in turn contributed to the monopoly of the water carriers. As a result, the emergence of a militant well guild, governmental disinterest in public utilities, and the powerlessness of local society provided an opening for the making of the water lords.

Within their water regime, by using a combination of customary law and physical force as needed, the water carriers imposed their power on a society that had been subordinated to them in consequence of their hegemony over the water supply. The carriers' water governance was so strong that they were called "water lords" by the residents of the neighborhoods they controlled. The water lords did not contribute to the construction of the urban community; rather, they usurped public space for their own needs to subjugate the state, piped water, and the city's residents. In this sense, water carriers gained extensive hegemony across territorial spaces for the purpose of extracting value. This would finally change in the decade that followed, when the water carriers and their regime came under the scrutiny of the nationalist government's political agenda.

CHAPTER 5

Hygienic Water: Down with the Water Lords, 1928–1937

In June of 1928 the Kuomintang (KMT) ended the Beiyang regime by capturing Beijing and moving the national capital to Nanjing (lit., "Southern Capital"). Beijing (lit., "Northern Capital") was renamed Beiping (lit., "Northern Peace") and was made the capital of Hebei Province until 1931. Compared to the Beiyang government, the Nationalist government of the KMT was more vigorous and vigilant in sustaining centralized power. Furthermore, the KMT's rebuilding of Beiping required the state to regain control of the water regime through a carefully coordinated sanitation campaign. Because the Nationalist government (1928–1949) gave more consideration to the health and lives of the people than to the carriers' needs, a constant and abundant supply of pure water was an object of vital importance to the new state. According to the state's water governance, it was unacceptable to leave the dispensing of such a necessity of life to the unlimited discretion of carriers who possessed an exclusive monopoly.

The state attempted to use the concept of public health, through which the city could be modernized and moralized, to harness water and discipline water carriers as part of its project of state-building. In other words, to control the water supply and force the carriers to come to terms with the authorities, the Nationalist government used the creation of hygienic water as an instrument of social manipulation, meanwhile, by decreasing the city's disease burden, hygienic water was also a vital means of strengthening the nation during the Nanjing decade (1928–1937). The state first tightened the grip on water carriers by outlining the regulations of their guilds, and then used hygienic campaign during the New Life Movement to reshape the water regime. Finally, the local municipal government and society tried to discipline carriers in dispute mediation.

Despite these efforts, the state did not have the ability to reform the water distribution territories, the bedrock of water power, and this inability anticipated the state's ultimate failure to overturn the water lords. In addition, the relocation of the national capital to Nanjing severely impacted Beiping's local economy. The city suffered from a tremendous loss of government funding and "an exodus of bureaucrats and their families, who had formed the wealthier

 | DOI:10.1163/9789004720800_007

part of the Beijing population."[1] This withdrawal of wealthy families led directly to the decline of the Waterworks, which relied on middle- and upper-class families as its customers. For instance, by 1933, piped households only numbered about ten thousand, accounting for just 3.3 percent of Beiping's total households; the remaining families depended on wells.[2] The decline of running water further highlighted the significance of wells and strengthened the water regime in post-1928 Beiping.

1 Reorganizing the Carriers' Association

On the eve of the arrival of the Nationalist government in December 1927, the Beijing municipal police department had already made plans to register all the carriers and outlaw water distribution territories.[3] The municipal police department issued twenty regulations concerning water carriers, including rules about registration, types of wells, the volume of buckets, and prices.[4] The Carriers' Association responded with strikes and a suspension of the water supply, but the police closely monitored the water carriers' movements after releasing its regulations. On December 14, the police broke up a meeting attended by more than thirty carriers at a teahouse. The next day, the police supervised another meeting called by the Carriers' Association.[5] On December 16, the police department issued a second order reaffirming its determination to weaken the water lords. Any carriers who increased water rates or engaged in disputes over water territories would be subject to severe punishment.[6]

Under pressure from the police department, the Carriers' Association pretended to comply with police regulations despite the water carriers' extreme opposition to such regulation. The police even threatened to confiscate water distribution territories. According to the water carriers, abolishing the territories would impose a huge loss on them since the territories were valued

1 Madeleine Yue Dong, *Republican Beijing: The City and Its Histories* (Berkeley: University of California Press, 2003), 106.

2 *Huabei ribao*, September 17, 1933. (華北日報 1933 年 9 月 17 日 第六版 全市水業受水閥把持，自來水用戶不及一萬。)

3 *Shuntien Times*, December 12, 1927. (順天時報 1927 年 12 月 12 日 第七版 警廳將定辦法，取締挑水夫。)

4 *Shuntien Times*, December 14, 1927. (順天時報 1927 年 12 月 14 日 第7版 警察廳打倒水閥。)

5 *Shuntien Times*, December 15, 1927. (順天時報 1927 年 12 月 15 日 第七版 取締水閥中井業公會今日開會。)

6 *Shuntien Times*, December 16, 1927. (順天時報 1927 年 12 月 16 日 第七版 警廳通令區署以嚴厲手段取締水閥。)

at dozens or hundreds of *yuan*. More essentially, they would lose the foundation of their monopoly. The Carriers' Association thus planned a meeting to discuss resistance but it was thwarted by the police.[7] The attitude of the police toward water distribution territories can be illustrated by a lawsuit that happened at the end of December when water carrier Zhang had a dispute with water carrier Lu over their territories. The court argued that the streets in fact belonged to the state, even though water carriers claimed the streets as private property and illegally monopolized the water supply. The court suggested abolishing water distribution territories to avoid further disputes.[8] On December 31, the municipal police department decided to outlaw 107 well houses and 2,702 water carriers through a code that would be put into effect on January 1, 1928.[9] Although, in the end, the code was never enforced because of regime change.

Water carriers were able to continue monopolizing the water supply after the KMT came to power, as exemplified in the following example: Water carrier Lu supplied water to the fourth police station located in his distribution territory. In return, the station provided him with two meals per day. The police station later relocated to a new site that was situated within carrier Er's territory, but it still used the service of carrier Lu. Carrier Er then took Lu's water wheelbarrow, and the police station had to turn to Er for drinking water.[10] If even the police could not choose their own water carrier, then certainly neither could ordinary residents. As one newspaper commented, water lords were more tyrannical than warlords, because if residents resisted water carriers' demands their water supply was suspended.[11] Like the warlords, the tyranny of water lords was finally curtailed because of the Nationalist revolution in 1928, although they would retain monopoly power over the water supply through the end of the Nationalist era.

To respond to the Nationalist revolution, on July 16, 1928, the water carriers converted their Carriers' Association into the Water Trade Union

7 *Shuntien Times*, December 21, 1927. (順天時報 1927 年 12 月 21 日 第七版 水閥仍積極保持權利。)

8 Beijing Municipal Archives, *Zhang Zhengzhou yu Lu Taigui yin zheng shuidao zhengchao yian*, J181-19-55781, December 1927. (北京市檔案館：張增壽與路臺桂因爭水道爭吵一案，J181-19-55781，1927 年 12 月。)

9 *Shuntien Times*, December 31, 1927. (順天時報 1927 年 12 月 31 日 第七版 警廳定明日起施行取締水糞兩業辦法。)

10 Beijing Municipal Archives, *Lu Ruijian bao E Shanglian qiangzhan shuidao*, J181-20-5882, March 1930. (北平特別市公安局 編存文卷，呂瑞儉報鄂尙連強占水道，請飭區保護，J181-20-5882，1930 年 3 月。)

11 *Shenbao*, April 9, 1928. (申報 1928 年 4 月 9 日 第 16 版 周素雅北京之三閥。)

(水行工會), a political labor union with its headquarters located in West Chashi Hutong (西茶食胡同). However, after the Shanghai Massacre of 1927, which aimed to expel the Chinese Communists and eliminate their influence in the revolution, the Nationalist government had launched a systematic campaign to control the longstanding labor movement. Indeed, to depoliticize the labor movement, in August 1928, the chief commander of the Beiping and Tianjin Garrisons, Yan Xishan (閻錫山, 1883–1960), ordered all unions to be dissolved. In response, the mayor of Beiping proposed gradually disbanding unions to avoid social unrest, and meanwhile dispatched armies to keep all unions under surveillance.[12] Although the order was not fully enforced, unions in Beiping were subject to rigorous screening. Thus, at its inception, the new Water Trade Union already faced heightened scrutiny.

An episode on Meishi Street (煤市街) demonstrates the Nationalist Party's interference in water carriers' affairs that ultimately led to the termination of the carriers' street protests and associated violence. On August 23, 1928, six waterworks employees who were installing plumbing on Meishi Street were attacked by dozens of water carriers. Four workers were injured and seventeen carriers were detained.[13] The Water Trade Union reported this clash to the General Union, the overarching organization established for all unions. The Civic Training Association of the KMT in Beiping asked the municipal government to seek a fundamental solution that would end the chronic disputes between the water carriers in the union and the running water workers.[14] The Water Trade Union even petitioned to dismantle the running-water pipes that were close to water distribution territories in order to maintain the water carriers' livelihoods.[15] But they soon felt an unprecedented pressure from the state.

The Nationalist government, which was growing more powerful, and indeed drifting toward authoritarianism, tightened its grip on society and refused to tolerate the street protests and social movements that had been a feature of city life since the mid-1920s. Focused on state building after revolution, the Nationalist Party and the government placed strict regulations on water carriers.

---

12 *Shijie ribao*, August 18, 1928. (世界日報 1928 年 8 月 18 日 第三版 平津各工會解散後當局正籌劃善後。)

13 *Shuntien Times*, August 24, 1928. (順天時報 1928 年 8 月 24 日第七版 前外煤市街南首水閥工人激戰。)

14 *Jingbao*, August 30, 1928. (京報 1928 年 8 月 30 日 第六版 北平水行工人與自來水公司之糾紛問題。)

15 *Shuntien Times*, December 14, 1928. (順天時報 1928 年 12 月 14 日 第七版 水閥與自來水 請撤接近水道水管。)

Somewhat cowed in the face of this vigorous party-state power, the water carriers restrained themselves and appeared subject to the party's authority. For example, in February 1929, the Beiping branch of the National Party convened water carriers from the Water Trade Union at Tiananmen Square to attend a meeting and a parade organized by the government.[16]

Under the vigilant surveillance of the KMT, water carriers had to resort to lawful and rational means to articulate their concerns. Henceforth, confrontations between carriers and the Waterworks were greatly reduced. Their last conflict took place in May 1929, when Waterworks workers were installing pipe at a private house in the western section of the city. Two dozen water carriers arrived to stop the installation and confiscate all the pipes and tools, but they did not beat and injure the Waterworks workers as they usually did. From this episode, we can observe that the violence that had characterized water carriers' street movements and protests in the mid-1920s was effectively terminated.[17] In addition, to weaken their solidarity, the KMT developed new tactics for dividing and supervising the carriers, based on the section of the city that a carrier served. For example, the KMT limited membership in the Water Trade Union to the western section of the city, while water carriers in the eastern section were not included; this accentuated the conflicting interests among water carriers and made it simpler to play them off against each other.

When a worsening drought in the summer of 1929 caused water shortages, water carriers raised water prices. These higher prices angered the municipal government, which prepared to do battle with the water lords and the Water Trade Union.[18] Meanwhile, to solidify their power, the water carriers appealed to the General Union in June 1929 to unite all the carriers in the city, and to reorganize the existing Water Trade Union (水行工會) into a Water Carrier Union (水夫工會) whose membership was open to all water carriers in the city.[19] With the supervision of the Civic Training Association of the KMT, the new Water Carrier Union was established in September, comprised of an executive committee with thirteen members and a supervising committee with five

16 *Shuntien Times*, February 25, 1929. (順天時報 1929 年 2 月 25 日 第七版 水夫因赴大會停止半日工作。)

17 *Shuntien Times*, May 14, 1929. (順天時報 1929 年 5 月 14 日 第七版 西城二龍坑丁字街水夫阻止裝安水管。)

18 *Shuntien Times*, June 6, 1929. (順天時報 1929 年 6 月 6 日 第 7 版 井商與水夫擡高水價。)

19 *Jingbao*, July 26, 1929. (京報 1929 年 7 月 26 日 第六版 水行工會改組 水夫工會將成立。)

members. Huang Qinghai was selected as the chairman of the union.[20] The municipal government levied on each well house an amount ranging from forty cents to one *yuan* per season, which contributed over one thousand *yuan* per year to the government's tax receipts.[21]

The new Water Carrier Union wielded formidable power over its members. Funds were collected based on water wheelbarrows, and members were charged ten cents per month per wheelbarrow. The member fee was mandatory and disobedience was not tolerated. One well-house owner, Xie, refused to pay dues. The union confiscated his two water wheelbarrows as a penalty, but Xie declined to pay the fine of six *yuan* to get the wheelbarrows back. A week later the union raided his well house and carried off two more wheelbarrows.[22] It employed the same practice to discipline well-house owner Gai who refused to pay his dues. The union ordered an executive committee member, Tian, to seize Gai's four water wheelbarrows as a punishment and this completely paralyzed his water business.[23]

The KMT's government would not allow the labor movement to remain independent of government control. In October 1929, a labor union law brought all labor union activity under governmental supervision. In 1930, the aggressive Nationalist government planned to thwart the labor movement by disbanding the General Union. Two representatives from the Water Carrier Union attended the General Union conference and opposed this order.[24] In response, the government brought the General Union under tighter control. In April 1931, the Hebei provincial government completed its reorganization of the General Union, licensed new organizations, and converted the Water Carrier Union back into a well guild. In other words, the Water Carrier Union was transformed from a political labor union into a pure business unit and was categorized as one of seventy-two merchant organizations in Hebei Province.[25]

---

20 *Huabei ribao*, September 20, 1929. (華北日報 1929 年 9 月 20 日 第六版 本市新聞 水夫總工會 昨開成立大會。)

21 *Beiping chenbao*, November 17, 1933. (北平晨報 1933 年 11 月 17 日 第六版 平市水閥調查。)

22 Beijing Municipal Archives, *Xie Qingrong konggao shuihang gonghui yin zhengshou huifei weisui jiang shuiche tuizou yian*, J181-21-3057, April 30, 1929. (北京市檔案館：解慶榮等控告水行工會因征收會費未遂將水車推走一案，J181-21-3057，1929 年 4 月 30 日。)

23 *Shuntien Times*, August 9, 1929. (順天時報 1929 年 8 月 9 日 第七版 法庭旁聽記 水閥的糾紛。)

24 *Shijie ribao*, March 1, 1930. (世界日報 1930 年 3 月 1 日 第六版 平市各工會昨開第五次代表大會。)

25 *Shijie ribao*, April 4, 1931. (世界日報 1931 年 4 月 4 日 第三版 冀人民團體審查將竣。)

The reorganization of the Water Carrier Union was just the beginning of subjugating the water carriers and their water regime. This disciplining of the Water Carrier Union cut off the carriers' connections with other unions and altered the parameters of the water carriers' power.

2 Creating Hygienic Water

2.1 *The Hygiene of Well Water*

In the first decade of the twentieth century, many city governments in China took the initiative to raise the level of public consciousness about health (*weisheng*) and shift the concept of sanitation from "guarding life" to what Ruth Rogaski has called "hygienic modernity," as an important aspect of modernizing the nation. Arriving along with imperialism in the late nineteenth century, the notion of hygienic modernity was a crucial element in the formulation of Chinese modernity; it espoused and encapsulated the ideas of national sovereignty, the cleanliness of bodies, and the fitness of races.[26]

After the 1911 Revolution, the newly founded Republican government continued the sanitation policies of the Qing period, although it did not pursue these policies with much rigor. It established several bureaus and departments of health that had a limited role in governance. For example, the departments of health for the Metropolitan Area and the Military Guard had only nominal powers as far as the city itself was concerned; they were limited to stamping regulations issued by the Sanitary Bureau of the Police Board. In addition to departments with such nominal duties, the sanitary department of the municipal council maintained a hospital for treating contagious diseases.[27] Moreover, the Sanitary Bureau of the Police Board was an office within the building of the head of police. Although certain regulations concerning sanitation were drawn up as police measures, the head office had only one or two clerks to perform the accompanying clerical work. Its functions, which merited a two-room office in the city headquarters, were diluted by the time they filtered down to the district level and were added to the duties and regular responsibilities of a district police officer, who had no extra help to guide him in these matters.

26 Ruth Rogaski, *Hygienic Modernity: Meanings of Health and Disease in Treaty-Port China* (Berkeley: University of California Press, 2004).

27 John B. Grant, *Report on a General Health Survey of Peking*, February 1922, RF RG 5 Series 2, SS-601, Box 55, F346, Rockefeller Archive Center, NEW YORK, 16.

Throughout its existence, the Sanitary Bureau was thus a very elementary sort of organization with no formally trained staff.[28]

Major urban changes were set in motion by the Nationalists at the end of the 1920s,[29] with the notion of hygienic modernity as a key concept used by the government to increase overall state control and to legitimize itself in the cities. Beginning in 1928, the Beiping municipal government increased public health campaigns and further reinforced them by establishing an intrusive police force. To apply the principles of cleanliness and orderliness in the city, the municipal government established the Public Health Bureau in 1928 to replace the former Sanitary Bureau. The new bureau was later incorporated into the Public Safety Bureau, but functioned independently with an annual fund of twenty thousand *yuan*.[30] In addition, the municipal government set up three public health demonstration areas, two in the Inner City and one in the Outer City, by emulating the Health Station run by the Department of Public Health at Peking Union Medical College. Each public health demonstration area served a population of over 100,000 people and provided a program in public health instruction and demonstrations of health-care delivery.

Drinking water was one of the primary targets of public health because of the city's notoriously brackish water. The new municipal government attempted to legitimize itself by supplying modern running water, but soon realized that although running water had steadily served about one-fifth of the population over the years, it was unable to replace well water in serving the entire city. As noted above, wells were in fact becoming increasingly indispensable in Beiping's city life, owing to the departure of elites from the city. More importantly, urban planning experts presented a new idea about the importance of well water. They suggested that running water, which relied exclusively on river water for its source, had been exposed to poison by bacterial bombs in air raids and was fatal to city dwellers. In contrast, wells were scattered across the city and were difficult to destroy in war.[31]

In view of their significance, wells and their carriers were subjected to the hygienic scrutiny of the new government. In 1928 the Ministry of Internal

28 "The Reminiscences of Doctor John B. Grant," *New York Times Oral History Program*, Columbia University Oral History Collection, Part III, no. 195, Medicine and Public Health (Glen Rock, NJ: Microfilm Corporation of America, 1976), 180–81.

29 Henrietta Harrison, *The Making of the Republican Citizen: Political Ceremonies and Symbols in China, 1911–1929* (New York: Oxford University Press, 2000).

30 Xu Duanqing, "Beiping zhi gongong weisheng yipie," *TUNG-CHI ACTA MEDICA* 4, no. 1 (1934): 72–73. (許端慶：北平之公共衛生一瞥，同濟醫學季刊第4卷第1期，1934年3月，72–73。)

31 Zhang Wu, *Zhengli Beijingshi jihuashu* (1928), 27–28. (張武著：整理北京市計劃書, 1928, 27–28。)

Affairs began to regulate drinking-water wells.[32] In Beiping, the director of the Public Health Bureau, Fang Yiji, an expert trained at the Public Health School of Harvard University, surveyed 485 wells that supplied 80 percent of the city's water. His results indicated that almost all the wells contained coliform bacteria.[33] Another medical research team later suggested that unhygienic well water was the cause of intestinal illness in Beiping, where the mortality rate was fifteen times higher than in Britain.[34] Wells were criticized not only from the viewpoint of water quality, but also from the standpoint of appearance. Unsightly wells, which impacted or threatened the townscape's value, would be eliminated to beautify the appearance of the city's streets.

In September 1928 the Public Works Bureau of the Beiping municipal government proposed examining drinking water for its adherence to the principles of sanitation.[35] The next year the Bureau of Public Administration together with the Public Health Bureau institutionalized a drinking water inspection system that would regularly assay the water and publicize its findings.[36] This policy set the standards for drinking water in Beiping, and hygienic scrutiny became routine in the following decade. The Public Health Bureau also scrutinized well covers, model wells, well disinfection, and water device sanitization to increase the state's control and establish the bureau's legitimacy.

The public health movement about drinking water in Beiping reached its peak during the national New Life Movement (新生活運動), which aimed to control Chinese lifestyles and social morality and was launched by the central government in 1934. Its principles are summarizable as two basic themes: cleanliness and discipline, which were considered the first steps in achieving a new life. As part of the movement, Beiping held annual city hygiene campaigns in parks, streets, and neighborhoods. A public health inspection squad was mobilized to inspect shops, hotels, restaurants, and other public facilities. Policemen also visited various districts to check on compliance.[37] The water

---

32 "Gaunli yinshuijing guize," *Neizheng gongbao* 1, no. 3 (1928): 39–40. (管理飲水井規則 部令第三九七號公布國民政府內政部內政公報 法規 1928, 1(3), 39–40。)

33 Fang Yiji, "Beipingshi zhi jingshui," *Weisheng yuekan* (March 1928): 7–10. (方頤積：北平市之井水，衛生月刊，1928 (3) ，7–10 工作報告。)

34 Yao Xunyuan, "Cangying jingshui lengyin yu Beipingren changweizheng zhi guanxi," *Yixue zhoukanji* 3 (1930): 58. (姚尋源：蒼蠅井水冷飲與北平人腸胃癥之關系，醫學周刊集 第 3 卷 1930，58。)

35 *Jingbao*, September 15, 1928. (京報 1928 年 9 月 15 日 第六版 公用事業發展的兩個步驟。)

36 Zhao Jianai, "Beiping tebieshi gongyongju 1929 niandu shizheng dagang," *Beijing dangan shiliao*, no. 1 (1992): 7–9. (趙家鼐選編：北平特別市公用局 1929 年度施政大綱，北京檔案史料 1992 年第 1 期, 7–9。)

37 Lillian Li, Alison Dray-Novey, and Haili Kong, *Beijing: From Imperial Capital to Olympic City* (Macmillan, 2008), 65.

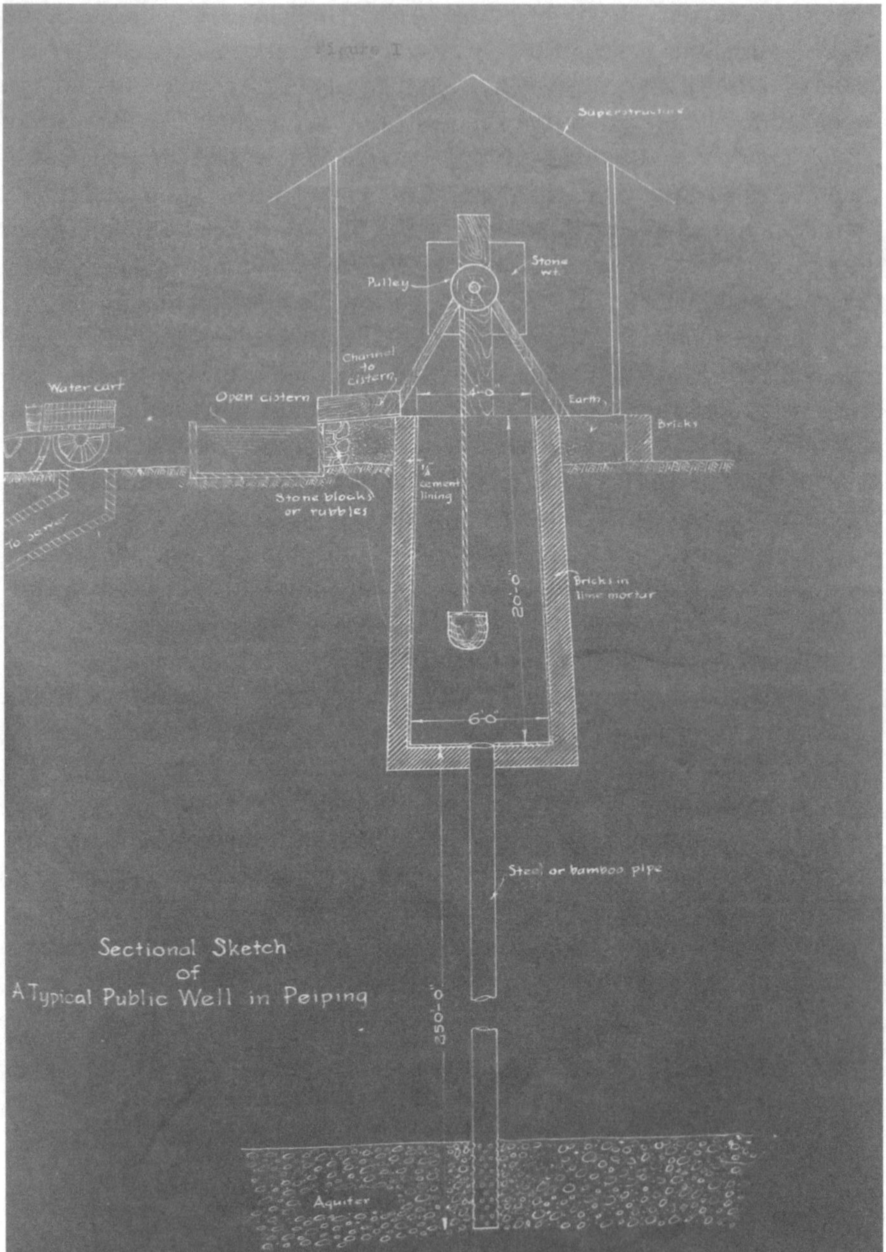

FIGURE 11 The sectional sketch of a well house with its well
SOURCE: EIGHTH ANNUAL REPORT OF THE DEPARTMENT OF PUBLIC HEALTH, JUNE 30, 1933, RG 5.3, SERIES 601J, BOX 220, F2740-2741, ROCKEFELLER ARCHIVE CENTER, NEW YORK

carriers and their water regime were put under full hygienic assessment and scrutinization.

2.2 *The Installation of Well Covers*

The first approach employed by the government to improve the drinking water in existing wells was to install covers to prevent ground dust and pollutants from entering them. Situated on the North China Plain, Beijing was notorious for its dusty winds to which wells without covers were constantly exposed. Covers were not only efficient in eliminating natural pollution but in preventing people and animals from falling in. Moreover, because of Beiping's impoverishment and rates of illness, the percentage of suicides was high, and this imposed another constant threat to wells.[38] It was not rare for a suicide to destroy a well. One young wife committed suicide in a sweet-water well whose owner employed more than ten water carriers to distribute water each day, but this suicide ruined the well.[39] Concerned with increasing contamination from suicides, residents petitioned the municipal government to install wooden fences.[40] In addition to human suicides, wandering animals also imperiled wells without covers. One public well without a cover was located at Xibian Gate and one summer morning a resident smelled an odor from the water; it turned out that a dog had fallen in.[41]

Despite these potential dangers, very few drinking-water wells in Beiping had covers. One survey conducted by the Sociology Department of Yenching University in 1929 indicated that the majority of wells on which Beiping residents relied had no covers and were prone to surface pollutants. As a result, the well water was intolerably filthy.[42] Huang Wanjie, a public health expert from National Beiping University, was able to demonstrate that well covers were an efficient means of preventing ground contamination. Huang found that a well in the Second Health Demonstration District was completely free from pollutants simply because it was sealed by a cover made of wood and iron costing 10 to 20 *yuan*.[43]

---

38 *Jingbao*, April 12, 1935. (京報 1935 年 4 月 12 日第六版。)

39 *Shijie ribao*, March 31, 1932. (世界日報 1932 年 3 月 31 日第八版 少婦負氣投井身亡 一處甜水井從此無人問津。)

40 *Shishi baihuabao*, March 15, 1930. (實事白話報 1930 年 3 月 15 日 各水井將安木柵欄。)

41 *Shuntien Times*, July 13, 1929. (順天時報 1929 年 7 月 13 日 第七版 西便內大道旁水井撈出死犬。)

42 Yu Xiezhong, "Beiping de gonggong weisheng," *Shehui xuejie, no.* 3 (1929): 64. (余協中：北平的公共衛生，社會學界 1929 年，第 3 期，第 64 页。)

43 Huang Wanjie, "Beiping yinshuijing zhi wuran laiyuan yuqi gaishan fangce," *Xinyiyao* 3, no. 9 (1935): 767–771. (黃萬傑：北平飲水井之汙染來源與其改善方策，新醫藥 1935 年第 3 卷 9 期, 767–71。)

A regulation enacted by the Ministry of Internal Affairs in 1928 specified that drinking-water wells should be equipped with covers.[44] In 1929, the Public Health Bureau of Beiping appealed to the Public Safety Bureau to order well owners to add wooden covers and lock them after sunset.[45] In May 1930, the municipal government announced that all wells in the city, whether privately or publicly owned, should have covers installed. It was estimated that the installation of covers for all governmental wells would cost one thousand *yuan*, but the government was unable to assign that amount of money.[46] Therefore, the policy for well covers was never fully implemented. In 1934, in response to a petition by a resident, the new mayor was still ordering the installation of well covers.[47] As late as 1937, one unsealed well remained near Tiananmen. It was constantly prone to contamination from suicides and was finally covered by the Public Works Bureau.[48]

While the installation of covers was a temporary solution for producing hygienic water by preventing external pollutants, unhygienic water also arose from within the wells themselves. One study indicated that both the walls and bottoms of the city's wells were lined with brick masonry over which a coat of cement-sand mortar was applied. Due to poor workmanship and temperature changes, however, most of the cement coatings became badly cracked within a few months. Well aprons, which were about 8–10 feet square and 2–3 feet high, were generally constructed with broken bricks or rubble or both, and then filled with earth, to serve mainly as working platforms. Shortly after construction, the center part of the aprons would become worn and slope towards the well walls. More importantly, since most wells were situated in congested parts of the city, there would usually be at least one private latrine formed from an earth pit or a brick-lined pit within 50 feet of the well.[49] To protect wells from such contamination, the municipal government strove to promote "model

44 "Gaunli yinshuijing guize," *Neizheng gongbao* 1, no. 3 (1928): 39–40. (管理飲水井規則 部令第三九七號公布國民政府內政部內政公報 法規 1928, 1(3), 39–40。)

45 *Shuntien Times*, November 17, 1929. (順天時報 1929 年 11 月 17 日 第七版 官私水井 均加木蓋 以免危險且重衛生。)

46 "Gongan mingling," *Beiping tebieshi shizheng gongbao*, no. 48 (June 1930), 2. (北平特別市市政公報 第四十八期 1930 年 6 月 公安命令 2。)

47 "Mingling," *Shizheng gongbao*, no. 252 (1934): 9–10. (市政公報 命令 1934 年 252 期 ，9–10。)

48 *Shijie ribao*, July 6, 1937. (世界日報 1937 年 7 月 6 日第五版 自殺無路 天安門玉帶河乾竭附近水井加添木蓋。)

49 "Comments on the Deep Wells in the First Inner Health District," *Eighth Annual Report of the Department of Public Health*, June 30, 1933, Appendix II, p. xiv, RG 5.3, Series 601J, Box 220, F2740-2741, Rockefeller Archive Center, New York.

wells" by constructing new wells and rebuilding old wells to meet sanitary standards.

2.3 *The Construction of the Model Well*

In designing its model well to solve hygienic problems, the government employed strict standards for well construction. In 1928 the Ministry of Internal Affairs issued national regulations for drinking-water wells and stipulated that the drilling methods, the well site, materials, the surrounding environment, and well depth should be thoroughly assessed. The regulations further specified the structure of model wells in order to prevent contamination, mandating that: (1) the wall of the well should be waterproofed to prevent percolation of sewage, (2) the depth of the well should be at least thirty feet, (3) the well should have a cover to prevent pollution, (4) the well curb should be two feet above the ground to prevent seepage from surface sewage, and (5) the site of well drilling should be at least 150 feet from latrines and sewers. Any violation by the well owner was subject to a punishment of a ten-day detention or a maximum penalty of ten *yuan*.[50]

During this period, the Public Health Bureau began scrutinizing wells in Beiping and outlawed unclean wells as a source of drinking water. One resident complained to the municipal government that a well on Fuyou Street (府右街) had been reopened for street sprinkling but had then been subleased to a merchant at three *yuan* per month for the sale of drinking water. The residents who used its water suffered from disease, and the Public Health Bureau shut down the well after its water was proved to be a danger to public health.[51] Meanwhile, the bureau ordered an assessment of all the wells in the city and forbid the use of polluted wells as a source of drinking water.[52] In 1929 the First Health Demonstration District surveyed its eighty-seven wells, including information on each well's owners, the diameter and depth of the wells, the structure of the well walls, construction materials, well covers, well curbs, water storage, and adjacent environments.[53]

In 1930 the Beiping municipal government issued its own version of regulations for drinking-water wells; these regulations granted the Public Health

---

50 "Gaunli yinshuijing guize," *Neizheng gongbao* 1, no. 3 (1928): 39–40. (管理飲水井規則 部令第三九七號公布國民政府內政部內政公報 法規 1928, 1(3), 39–40。)

51 "Gongan mingling," *Beiping tebieshi shizheng gongbao*, no. 8 (August 1929): 4. (北平特別市市政公報 第八期 1929 年 8 月 衛生文電 4。)

52 *Huabei ribao*, August 14, 1929. (華北日報 1929 年 8 月 14 日 第六版 衛生局禁售汙井水必要時或一律封鎖。)

53 *Beiping tebieshi diyi weisheng shiwusuo disinian nianbao* (Beiping, 1929), 24–27. (北平特別市第一衛生事務所第四年年報 1929 年 24–27。)

Bureau the authority to license well drilling, examine well water, and close down unqualified wells. The regulations also stipulated that the well owner had to submit an application to the Public Health Bureau and Public Works Bureau for well drilling that included details regarding the well site; the purpose of the well; the drilling method; the depth, diameter, and material of the well wall and cover; and the height of the well curb.[54] The Public Safety Bureau also launched a campaign against unhygienic wells by surveying all the wells in the city.[55]

Based on the fact that well drilling was so closely related to the construction of model wells, the Public Works Bureau ordered all well-drilling companies to register with the Social Bureau and the Public Works Bureau.[56] At that time there were seventeen well-drilling companies in Beiping. The most prominent of these was the Chinese Well Drilling Company located in Xidan, which drilled wells for stores, households, government buildings, and schools.[57] The remainder were small companies whose capital ranged from forty to two hundred *yuan*. However, the actual cost for drilling a model well was over five hundred *yuan*, so it was necessary to ask well-drilling companies to increase their capital to five hundred *yuan* before they could register with the city.[58] In addition, the Public Health Bureau explained to well drillers the model-well standards and instructions that they were expected to follow.[59]

The scrutiny of wells reached its height during the New Life Movement in 1934. The director of the Public Health Bureau, Fang Yiji, advocated improving drinking water by having well construction follow the standards for the model well.[60] In July 1934, the Public Health Bureau revised its regulations on drinking-water wells to include an elaboration of the standards for model well construction in Beiping. For example, the depth of the well had to be at least thirty feet; the well curb had to be at least two feet above the ground; the well wall had to be waterproof; the well had to have a cover or well shelter; the water

54 "Beipingshi yinshuijing qudi guize," 1929, *Beipingshi shizheng fagui huibian* (Beiping, 1934), 44. (北平市飲水井取締規則，1929 年，北平市政府參事室編：《北平市市政法規匯編》1934 年，44。)

55 *Huaibei ribao*, April 2, 1931. (華北日報 1931 年 4 月 2 日 第六版 整理水井計劃。)

56 *Shijie ribao*, August 17, 1932. (世界日報 1932 年 8 月 17 日第八版 鑿井商速註冊。)

57 *Beiping chenbao*, April 5, 1936. (北平晨報 1937 年 4 月 5 日 第二版 鑿井可致巨富。)

58 *Jingbao*, August 17, 1932. (京報 1932 年 8 月 17 日 第 6 版 工務局註意鑿井工程。)

59 *Huabei ribao*, September 12, 1935. (華北日報 1935 年 9 月 12 日 第六版 衛生局改善鑿井方法。)

60 *Huabei ribao*, March 28, 1934. (華北日報 1934 年 3 月 28 日 （新生活）水進行整頓。)

tank had to be above the ground with covers and a drain hole; and the well had to have drainage to drain off the water around it.[61]

The Public Health Bureau acted in accordance with hygienic principles to recognize and license wells, stating that newly drilled wells had to strictly conform to the bureau's well standards and old wells had to be improved or would otherwise be shut down.[62] One factor that contributed to the lack of sanitation of Beiping's wells was public latrines. A survey of public toilets in Beiping revealed that only 40 out of 570 drinking water wells were located more than forty feet from public toilets, which was the minimum safe distance for sanitation according to public health experts.[63] One drinking water well in Dagouxiang (大溝巷) was only thirty feet away from a shabby lavatory, while another well at Baiheyuan (百合園) was only twenty feet away from a public latrine.[64] These wells were polluted by coliform bacteria that were harmful to public health. The government ordered that either the wells had to be reconstructed or the latrines had to be moved. As another example, in the Second Health Demonstration District, where four wells were located close to latrines, the Public Health Bureau dismantled and relocated three of these toilets in 1934.[65] While a well in Chunshu Hutong (椿樹胡同) that was adjacent to a public toilet was relocated to protect the drinking water.[66]

In January 1935, the Public Health Bureau again amended its regulations on drinking-water wells and imposed new restrictions.[67] The bureau ordered all drinking-water wells to be registered, including the well's location, structure, water-drawing devices, water-storage facility, drainage facility, well house, and the purpose of the well.[68] The case of a well on Huayan Street (華嚴路) illustrates the execution of this policy. The well owner, Huang, had a sweet-water well drilled in 1917. At the request of nearby residents, Huang hired a

61 *Huabei ribao*, July 9, 1934. (華北日報 1934 年 7 月 9 日 第八版 飲水井管理規則。)

62 *Beiping chenbao*, July 26, 1935. (北平晨报 1935 年 7 月 26 日 第六版 卫生局拟定整顿水井计划。)

63 "1934nian Beipingshi zhengdun gongce shiliao," *Beijing dangan shiliao*, no. 3 (2007): 5–10. (1934年北平市整頓公廁史料，北京檔案史料 2007 年第 3 期, 5–10。)

64 "1934nian Beipingshi zhengdun gongce shiliao," *Beijing dangan shiliao*, no. 3 (2007): 24–29. (1934年北平市整頓公廁史料，北京檔案史料 2007 年第 3 期, 24–29。)

65 *Beipingshi weishengchu dier weishengqu shiwusuo diyi niandu nianbao* (Beijing, 1934), 35–37. (北平市衛生處第二衛生事務所編：北平市衛生處第二衛生區事務所第一年度年報 1934 年, 35–37.)

66 "Zizhi wendian," *Beiping tebieshi shizheng gongbao*, no. 18 (November 1929):5. (北平特別市市政公報 1929 年 11 月 第十八期 自治文電 5。)

67 *Shijie ribao*, January 13, 1935. (世界日報 1935 年 1 月 13 日第八版 平市府昨公佈飲水井取締規則。)

68 *Huabei ribao*, February 11, 1935. (華北日報 1935 年 2 月 11 日 第六版 舉行飲水井登記。)

FIGURE 12 A well house (right) adjoining a public toilet (left) in Beijing
SOURCE: SYDNEY D. GAMBLE PHOTOGRAPHS COLLECTION, DUKE UNIVERSITY LIBRARY

well keeper to watch over the well and charge for drawing water, but the well was not licensed to sell water. In November 1935 an anonymous letter accused Huang of selling unhygienic water without the permission of the Public Health Bureau. The bureau dispatched an inspector to investigate and found that the brick and cement construction of the well was good, the only offense was three fish in the water trough. However, Huang was out of town, and it was unclear whether or not the well was registered with the bureau. Huang immediately submitted a license application when notified by the bureau, which then sent the inspector again to verify that the structure and adjacent environment of the well were good. Accordingly, the chief of inspection decided that the well could be allowed to supply water to residents, but it had to be under hygienic scrutiny as an operating well. After Huang agreed to install a pump, add a cover, and build a curb and storage facility, the Public Health Bureau licensed the well.[69]

69 "Minguo shiqi Beipingshi gongyong yinshuijing guanli shiliaoxuan," *Beijing dangan shiliao*, no. 4 (2006): 56–68. (民國時期北平市公用飲水井管理史料選，北京檔案史料，2006 年第 4 期，58–68。)

The New Life Movement also promoted the construction of model wells. For example, the Third Health Demonstration District reconstructed six model wells in 1936.[70] However, the cost of each model well ranged from 400 to 500 *yuan,* which impeded the progress of their construction.[71] The well owners were unwilling and unable to pay this price, while the Public Health Bureau was short of funding. One suggested solution was that the Public Health Bureau raise funds through loans and lend the money to well owners for reconstruction.[72] Nevertheless, this idea was never put into practice, and the high costs constrained municipal governments' construction of model wells. Even though the model well program was initiated as early as 1930, Fang Yiji, the Public Health Bureau's director, was still requesting that model wells be built in Beiping in December 1935.[73]

2.4 *The Disinfection of Wells*

Beiping residents had long boiled their drinking water because, according to the theory of traditional Chinese medicine, hot water could be used to expel excess cold from the body. This widespread practice reduced toxins, yet boiling was unable to sterilize all water-borne bacteria and heat-stable coliform bacteria remained a pressing problem. As early as 1928, Fang Yiji suggested, that since the permanent solution of reconstructing wells was impractical given the cost, a temporary solution for eliminating coliform bacteria was disinfection. The method was to sanitize well water with a 1:10 dilution of chlorinated lime, which would neutralize the bacteria within thirty minutes and remain effective for five to twelve hours, depending on the well's structure.[74] However, since neither the government nor society showed much enthusiasm for a disinfection campaign, it was not enforced until 1932 when the fear of poison played a significant role in promoting it.

---

70 *Beipingshi weishengju disan weishengqu shiwusuo diernian nianbao* (Beijing, 1936), 33–39. (北平市衛生局第三衛生區事務所編：北平市衛生局第三衛生區事務所第二年年報 1936 年，33–39。)

71 Huang Wanjie, "Beiping yinshuijing zhi wuran laiyuan yuqi gaishan fangce," *Xinyiyao*, 3 no. 9 (1935): 767–771. (黃萬傑：北平飲水井之汙染來源與其改善方策，新醫藥 1935 年第 3 卷 9 期, 767–771。)

72 *Beiping chenbao*, September 10, 1934. (北平晨報 1934 年 9 月 10 日 第十三版 衛生月刊。)

73 *Huabei ribao*, December 18, 1935. (華北日報 1935 年 12 月 18 日 第六版 衛生局註重全市飲水清潔。)

74 Fang Yiji, "Beipingshi zhi jingshui," *Weisheng yuekan* (March 1928): 7–10. (方頤積：北平市之井水，衛生月刊，1928 (3) ，7–10 工作報告。)

Although disinfection was unrelated to the possibility of someone poisoning a well, episodes of unproved well poisoning in 1932 heightened concerns about the hygiene and safety of drinking water. This finally motivated both the government and the general public to promote the importance of water cleanliness. Sino-Japanese relations had intensified after the Japanese occupied Manchuria in 1931, and North China, including its frontier city Beiping, had fallen under the Japanese military shadow. It was said that more than one thousand accomplices to Japan were located throughout Hebei Province. In May 1932, the police in Hebei Province arrested several Shandong men who were collaborators hired by the Japanese at three *yuan* per day to poison wells. The poison they carried was extremely toxic and water-soluble; it caused symptoms of swelling eyes and throats, and would kill people in a few days.[75] Throughout May it was reported that at least six counties in Hebei Province had found traitors who poisoned wells and caused great social panic. For example, one village in Daxing County (大興縣) south of Beiping witnessed the death of thirteen persons in three days. All of the deceased had drunk water from one well in the village, in which the county police found poison.[76]

Incidents of well poisoning finally reached Beiping. On June 1, a man from Hebei Province approached a well outside Desheng Gate (德勝門) and poured something into the well. Fortunately, it was discovered by the owner's daughter and neighbor, and they cooperated with the police who caught the suspect. The police found an iron box with three pills inside the well. As a consequence, the municipal government was ordered to keep a vigilant eye on wells.[77] In the following days, several cases of well poisoning were reported in the western part of Beiping.[78] The well poisonings terrified Beiping's residents; one newspaper even published the Chinese medical antidote to poison.[79] The mass panic also caused misunderstandings. On June 7, a pedestrian Zhang and his nephew approached well owner Lu's water tank to quench their thirst. Lu was very suspicious of strangers after learning about a case of well poisoning from the newspaper. When Lu noticed Zhang pocketing a black medicine bottle,

75 *Shijie ribao*, May 8, 1932. (世界日報 1932 年 5 月 8 日第八版 井中投毒藥 槁城縣捕獲漢奸。)

76 *Shijie ribao*, May 21, 1932. (世界日報 1932 年 5 月 21 日 第八版 平南劉家莊發現井中投毒。)

77 *Shijie ribao*, June 2, 1932. (世界日報 1932 年 6 月 2 日 第八版 昨日德勝門外捕獲井中投毒匪徒。)

78 Shijie ribao, June 5, 1932. (世界日報 1932 年 6 月 5 日第八版 井中投毒漢奸。)

79 *Shijie ribao*, June 8, 1932. (世界日報 1932 年 6 月 8 日第八版 誤服井中毒水，救急方法 中藥方子。)

he called the police who arrested Zhang. It turned out that the medicine the innocent Zhang carried was used to cure hemorrhoids.[80] Two days later Yang, an apprentice in a car shop, went to see a doctor. With his medicine in his hand, he refreshed himself in the shade of a tree near a foreign well. A soldier from the garrison headquarters who was watering his horse thought that Yang was trying to poison the well, and took him to the police.[81]

Although cases of well poisoning were repeatedly reported in the newspaper, details about how the collaborators were punished or how to prevent well poisonings was never provided, as one commentator noted in an editorial.[82] Nonetheless, the widely circulated rumor of well poisonings in 1932 raised public awareness about the cleanliness and security of drinking water, and this provided the impetus to disinfect wells. Meanwhile, disease in the city also highlighted the importance of disinfection. Since the 1920s, Beiping had been afflicted with dysentery and cholera. In July 1932, the Plague Prevention Committee decided to allot five thousand *yuan* in funding for well disinfection.[83] The Public Safety Bureau, in association with the Public Health Bureau, was in charge of this new campaign. First, the police together with a sanitation officer would inspect and examine wells once a week; second, the sanitation officer would train the police inspector regarding disinfection; third, the police department would explain and disseminate well disinfection and the chlorinated liquid it used.[84] To promote disinfection, the Plague Prevention Committee also distributed a documentary on cholera to be played in the movie theaters of the city.[85]

80 Beijing Municipal Archives, *Jiesong Lu Fugui kong Zhangshan xieyou heise yaoshuiping zai yi suokanguan shuijingpang heshui douliu buqu*, J181-21-15499, June 7, 1932. (北京市檔案館：解送路福貴控張山等攜有黑色藥水瓶等物在伊所看管水井旁喝水逗留不去J181-21-15499，1932年6月7日。)

81 Beijing Municipal Archives, *Yang Zibin xiadu an*, J183-2-6762, June 9, 1932. (北京市檔案館：楊子彬下毒案，J183-2-6762，1932年6月9日。)

82 *Shijie ribao*, June 12, 1932. (世界日報 1932年6月12日第八版 由井中投毒說到賞罰問題。) Only one case of well poisoning was verified in 1934. The sixty-seventh regiment stationed at Langfang (廊坊) arrested a Chinese collaborator who had been bribed by the Japanese for thirty yuan to poison wells; he was executed four days later. See *Jingbao*, January 21, 1934. (京報 1934年1月21日 第6版 驚人大毒計。); *Jingbao*, January 25, 1934. (京報 1934年1月25日 第6版 向防地水井置毒的漢奸張永明將執行死刑。)

83 *Shijie ribao*, July 15, 1932. (世界日報 1932年7月15日第八版 平市防疫費定爲五千元將開始井水消毒工作。)

84 *Beiping chenbao*, July 16, 1932. (北平晨報 1932年7月16日 第六版 註重水井消毒。)

85 *Shijie ribao*, July 29, 1932. (世界日報 1932年7月29日 第八版 防疫委員會工作忙 霍亂影片分送各影院映演。)

The poisonings led to well disinfection, but also caused a misconception. The disinfectant used was chlorine, which proved effective in eliminating bacteria. However, it had a medicinal taste and tainted the water with an odor that reminded residents of poisoned wells. Thus, the suspicious well owners tended to misunderstand the significance of disinfection and thought the medicine poured into their wells made the water taste like the water from wells poisoned by Chinese collaborators; they were thus reluctant to subject their wells to disinfection.[86] The well owners prioritized taste over health, while the state prioritized health over taste. The authorities had to emphasize the benefits of disinfection and clarify that the chlorine was harmless.[87] The newspapers also encouraged residents to appreciate and cooperate with the Plague Prevention Committee in enforcing well sanitization in order to prevent pestilence.[88] In September 1932, the Plague Prevention Committee decided to institutionalize well disinfection and scheduled it from April to September each year. From that time forward, well disinfection became routine in Beiping.[89] During the campaign of 1932, the First Health Demonstration District sanitized its thirty-four wells sixty-three times each.[90]

All wells were chlorinated once every several days by sanitation policemen. A chlorine solution made of bleaching powder was prepared and distributed every Sunday to each well and stored there in a dark place. The policeman, equipped with a sounding rope, a measuring glass, and a dosage chart went in the morning from one well to another, measured the depth of the water, looked at the chart to ascertain the required dosage, obtained the solution from the well owner, dropped it in the well, and entered the measurements and his signature into a record that was also kept by the well owner. This arrangement was not only time-saving but also obviated the danger from over-dosing. The cost of chlorination was about seven cents per ten wells per day, excluding labor.[91]

---

86 Jie, "Jingshui xiaodu de yiyi," *Beiping yikan* 2, no. 9 (1934): 35–36. (傑：井水消毒的意義 北平醫刊 1934 年 10 月 第 2 卷第 9 期，第 35–36 页。)

87 *Jingbao*, August 4, 1932. (京報 1932 年 8 月 4 日 第 6 版 井中灑入氯氣液 微帶藥性，飲之有益。)

88 *Jingbao*, August 28, 1932. (京報 1932 年 8 月 28 日 第 6 版 小評二 井水消毒。)

89 *Shijie ribao*, September 25, 1932. (世界日報 1932 年 9 月 25 日第八版 井水多含腸菌 四月至九月施行消毒。)

90 *Beiping tebieshi diyi weisheng shiwusuo diqinian nianbao* (Beiping, 1932), 35. (北平特別市第一衛生事務所第七年年報 1932 年， 35。)

91 "Public Well Water," Eighth Annual Report of the Department of Public Health, June 30, 1933, p. 16, RG 5.3, Series 601J, Box 220, F2740-2741, Rockfeller Archive Center, New York.

FIGURE 13 Sanitation policeman with chlorination box
SOURCE: EIGHTH ANNUAL REPORT OF THE DEPARTMENT OF PUBLIC HEALTH, JUNE 30, 1933, RG 5.3, SERIES 601J, BOX 220, F2740-2741, ROCKEFELLER ARCHIVE CENTER, NEW YORK

During the New Life Movement in 1934, the Public Health Bureau assigned twelve inspectors to sanitize each well once a day, and they examined well water 747 times.[92] The Second Health Demonstration District conducted 1,782 disinfections and each well was disinfected 52 times from May to August. The inspector posted a form in each well house with information about the date, well depth, the usage of coliform lime, and the signature of the inspector.[93] However, this disinfection regimen was a heavy duty for twelve inspectors, so in 1936 the Public Health Bureau adjusted its policy, ordering well owners to sanitize their wells twice a day from June to September with free coliform lime distributed by the bureau. The bureau would conduct random sample inspections and any offender would be subjected to heavy penalties.[94] Along with disinfection, well owners were cautioned to keep the surface of the apron and the surroundings clean and to avoid surface pollution as much as possible.

2.5 *Sanitizing the Water Carriers*

The construction of the model well and well disinfection ensured the production of hygienic water. However, drinking water was supplied and distributed by water carriers, who played an important role in producing hygienic water through drawing, storing, and delivering it. At any of these stages, contaminants could be introduced. So while wells were one of the main sources of pollution, the water carriers were another. For verification of contamination by carriers' hands, the Department of Public Health at PUMC collected a sample of water immediately after the well had been chlorinated and the sample had been transferred from one bucket to another. It was striking that a dramatic increase of bacteria in one cc of water could result from several contacts with the hands of a water carrier (Table 6). Thus, the Department concluded that even when a well was adequately protected and disinfected there was still the danger of it being reinfected through the unsanitary delivery procedure.[95] Accordingly, water carriers and their facilities were placed under rigid review.

---

92 Beipingshi zhengfu weishengju, *Beipingshi zhengfu weishengchu yewu baogao* (Beiping, 1934), 50–52. (北平市政府衛生局編印：北平市政府衛生處業務報告 1934 年，50–52。)

93 *Beipingshi dier weisheng shiwusuo dier niandu nianbao* (Beiping, 1935), 41–45. (北平市衛生處第二衛生區事務所編：北平市衛生處第二衛生區事務所第二年度年報 1935 年, 41–45。)

94 *Shijie ribao*, June 4, 1936. (世界日報 1936 年 6 月 4 日第六版 平市井水消毒 定十五日開始實行。)

95 "Comments on the Deep Wells in the First Inner Health District," *Eighth Annual Report of the Department of Public Health*, June 30, 1933, p. XVII–XIX, RG 5.3, Series 601J, Box 220, F2740-2741, Rockfeller Archive Center, New York.

TABLE 6 Extent of contamination of water by hands of water carrier

Sample	No. of bacteria per cc found in sample				Date of exam
	Well	Open Cistern	Water Cart	Pail	
No. 1	10	116	440	--	May 29, 1933

In 1928 the First Health Demonstration District investigated the sanitation of water storage cisterns, buckets, carrying poles, water carriers, and their boarding practices in well houses.[96] Under the name of public sanitation, in 1930, the Public Health Bureau issued regulations for water carriers that specified rules they had to comply with: (1) water carriers should report and register their name, age, birthplace, address, and water distribution territories to the Public Works Department in order to obtain a license; (2) buckets should adhere to a standard bucket size; (3) the price of water should be fixed; (4) water carriers could not extort or meddle with customers or suspend their water supply; (5) water carriers should pay attention to the sanitation of water and water devices; (6) water carriers could not prevent the installation of running water; and (7) violators would be subjected to a maximum penalty of five *yuan* or have their license suspended, while those who were convicted but refused to pay the fine would be placed in custody.[97] As these regulations made clear, the goal of the regulations went beyond considerations of hygiene to include fair sales practices. In the face of strong state intervention, the well guild even disciplined itself. It informed the Social Bureau that, because some shortsighted well-house owners had connived with their carriers to monopolize the water supply and increase water prices, the well guild had decided to standardize the bucket size and set fixed water rates.[98]

During the New Life Movement, water carriers and their devices were subjected to strict inspection. Water carriers drew water from wells and stored it in a water tank before loading it into water carts. This water tank was located

96 *Beiping tebieshi diyi weisheng shiwusuo disinian nianbao* (Beiping, 1929), 24–27. (北平特別市第一衛生事務所第四年年報 1929 年 24–27。)

97 "Beipingshi shoushuifu guanli guize (1930)," *Beipingshi shizheng fagui huibian* (Beiping, 1934), 55–56. (北平市售水夫管理規則 1930 年，北平市政府參事室編：北平市市政法規匯編 1934 年，55–56。)

98 *Shijie ribao*, November 28, 1931. (世界日報 1931 年 11 月 28 日 第八版 打倒水閥。)

beside the well and was easily exposed to external pollutants, so it was the first target for sanitation. For example, water carrier Wang ran a water business on Andingmen Street. The sanitation inspector detected that Wang's water tank had sand and mud in it, so Wang was fined one *yuan* and ordered to clean the tank.[99] Well owner Dong was fined thirty cents because of his dirty water tank.[100] Water carrier Zhao was fined one *yuan* because of a filthy water tank.[101] Another Shandong water carrier, Jiang, whose cistern was unclean was fined thirty cents.[102]

Water wheelbarrows were the second item screened by public health officers. The wheelbarrows that water carriers used to distribute water were considered outdated and did not conform to the sanitation standards. Technicians from the Public Health Bureau and the Public Utilities Bureau planned to design a new water cart at the water carriers' expense to make sure the water being delivered was clean.[103] In 1934 the Public Health Bureau reemphasized the importance of designing water wheelbarrows to prevent water carriers' dirty hands from contacting and tainting the drinking water they delivered.[104] Together with these devices, water carriers' bodies were also redefined and secured according to hygienic standards. Water carriers' hands and feet were considered a medium of water contamination when they stood at the well curb to draw water in buckets by hand.[105] Moreover, water carriers' daily living habits, such as boarding and toileting practices, were also subjected to hygienic scrutiny.[106]

---

99 Beijing Municipal Archives, *Chahuo Wang Jichen yin shuicao bujie yinshui youai yian*, J183-2-7595, June 25, 1934. (北京市檔案館：查獲王紀臣因水槽不潔飲水有礙一案，J183-2-7595，1934 年 6 月 25 日。)

100 Beijing Municipal Archives, *Chahuo Dongwangshi kaishe shuijing yingsheng, shuicao bujie yian*, J183-002-08925, April 6, 1935. (北京市檔案館：查獲董王氏開設水井營生，水槽內不潔一案, J183-002-08925, 1935 年 4 月 6 日。)

101 Beijing Municipal Archives, *Chahuo Zhao Rongsheng yin shuicao nei bujie yian*, J183-002-08961, May 27, 1935. (北京市檔案館：查獲趙榮盛因水槽內不潔一案, J183-002-08961, 1935 年 5 月 27 日。)

102 Beijing Municipal Archives, *Jiang Chuanyi bushua shuicao yian*, J184-002-17306, September 5, 1935. (北京市檔案館：蔣傳義不刷水槽一案, J184-002-17306, 1935 年 9 月 5 日。)

103 Zhao Jianai, *Beiping tebieshi gongyongju 1929 niandu shizheng dagang*, Beijing dangan shiliao, no. 1 (1992): 9. (趙家鼐選編：北平特別市公用局 1929 年度施政大綱，北京檔案史料 1992 年第 1 期，9。)

104 *Huabei ribao*, March 28, 1934. (華北日報 1934 年 3 月 28 日 （新生活）水進行整頓。)

105 *Beiping chenbao*, Feburary 15, 1934. (北平晨報 1934年2月15日 第十一版 衛生月刊。)

106 Beipingshi zhengfu weishengju, *Beipingshi zhengfu weishengchu yewu baogao* (Beiping, 1934), 36–38. (北平市政府衛生局編印：北平市政府衛生處業務報告 1934 年，36–38。)

To conclude, in boosting morality and health, the state formulated its own solution as an alternative form of water governance. The municipal government launched hygienic campaigns by instituting various forms of regulation related to hygiene. By engineering public health, the government attempted to remake the city through the control of water and its carriers. Yet, the hygienic disciplining of the water carriers did not address water distribution territories, which foreshadowed further actions that would be taken against the water lords.

3 Down with the Water Lords: The Municipal Government and the Carriers

The Nationalist government employed the discourse of hygienic modernity as an approach to improving water governance controlling the water supply. These intensive hygienic campaigns guaranteed access to clean water but did not destroy the water regime. The ambitious nationalist state gradually recognized that supplying water to the populace was a primary duty of government. It thus sought to remove water from the hands of private carriers and receive income from water sales to strengthen the local government. As early as 1933, Mayor Yuan Liang, who had a profound awareness of the water problem because of his long residence in Beiping, proposed outlawing water carriers and nationalizing the water supply.[107] When Yuan Liang began to address the excess power of the water lords, Xuan used this opportunity to appeal to the Police Bureau. Under pressure from the police, Huang promised to not obstruct Xuan's business, but later Xuan was again stopped by Huang.[108] The desperate Xuan had to suspend his business. The case of Lama Temple alludes to the inability of the state to tackle the water lords; it also shows that the carriers were determined to guard their territory. Accordingly, the municipal government initiated another campaign against the water regime.

Since water carriers still monopolized the water supply and constantly squeezed residents by increasing water prices and reducing water volume, the Public Health Bureau continued to promote a standard size for buckets and to control changes in water prices by setting a fixed price for all carriers.

107 *Shijie ribao*, August 16, 1933. (世界日報 1933 年 8 月 16 日 第八版 袁良昨談市政。)

108 Beijing Municipal Archives, *Beipingshi gonganju guanyu shuifa Huang Qingshan lvci lanzu yingye ouda gongyou yian*, J183-002-07463, November 20, 1933. (北京市檔案館：北平市公安局關於水閥黃慶山等屢次攔阻營業毆打工友一案，J183-002-07463, 1933 年 11 月 20 日。)

In 1935 the municipal government standardized measurements by burning illegal measuring instruments in front of Tiananmen and setting the capacity of a standard bucket at fifteen kilos.[109] The well guild played an important role in protecting the water regime by disciplining its own carriers and fending off the interventions of the state. Until 1934, a total of 108 well houses, which accounted for two-thirds of the city's well houses, were members of the guild.[110] The well guild finally responded to the government's call for standardized measurements one year later and petitioned to fix water prices and standardize the bucket size themselves in order to mitigate pressure from the authorities. The Public Health Bureau arranged a meeting with the well guild and stipulated that the price of delivered water would be ten cents per fourteen buckets and self-delivered water would be ten cents per thirty-six buckets, while the capacity of the bucket would be fifteen kilos.[111] The Public Health Bureau also planned to make two thousand registration plates for water wheelbarrows and to distribute working permits to water carriers to facilitate inspection following the regulation of water carriers.[112]

To prepare the ground to ban the carriers, the Public Health Bureau also started to register all water carriers.[113] In May 1935 the bureau revised its regulations and ordered water carriers to register.[114] The bureau convened 137 well-house owners and the chief inspector explained the regulations to them, specifically elaborating the rules pertaining to water carriers' extortion of residents.[115] In March 1936 the Public Health Bureau was finally determined to enforce the registration of all well houses, including well owners, water carriers, and water wheelbarrows. It also encouraged residents to report water carriers' misconduct, such as price gouging and suspended water delivery.[116] However, only forty well owners and a dozen water carriers out of one thousand completed registration by April. The rest remained reluctant to register

109 *Shijie ribao*, January 26, 1935. (世界日報 1935 年 1 月 26 日 第八版 不合法之度量衡舊器抵定期在天安門焚燬。)

110 *Yishibao*, July 23, 1934. (天津益世报 1934 年 7 月 23 日 北平的井业。)

111 *Jingbao*, February 21, 1936. (京報 1936 年 2 月 21 日第六版 社會衛生兩局會同劃一全市井水價格。)

112 *Shijie ribao*, March 3, 1936. (世界日報 1936 年 3 月 3 日第六版 衛生局管理售水夫。)

113 *Beiping chenbao*, January 30, 1935. (北平晨報 1935 年 1 月 30 日 第六版 水閥亦將取締先登記水井入手。)

114 *Jingbao*, May 23, 1935. (京報 1935 年 5 月 23 日 第六版 爲便利市民起見，衛生局即取締水閥。)

115 *Huabei ribao*, August 16, 1935. (華北日報 1935 年 8 月 16 日 第六版 衛生局朱鼎鈞昨向全市水夫訓話。)

116 *Huabei ribao*, March 18, 1936. (華北日報 1936 年 3 月 18 日 第六版 衛生局整頓全市飲水 舉辦售水夫登記。)

because they were afraid the government was planning to reorganize the water distribution territories. After the well guild submitted a petition, the bureau postponed registration.[117]

To consolidate its regime, the guild enacted and enforced rules and penalties against nonmembers and members who refused to pay the trade dues and fees.[118] The well guild also got involved in the internal conflicts between water carriers and well-house owners. One owner, Xie, ran a water business with three wells and a large water distribution territory. Xie subleased his territory to five carriers, one of whom stole the contract, then conspired with the other four lessees to drill a new well and claim ownership of the leased territory. This led to a conflict with Xie, who finally regained his territory with the guild's interference.[119] The well guild was professional in addressing such conflicts. For example, the aunt of water carrier Jiang subleased her territory to Wang for fifteen years. When the lease expired, the lessee Wang refused to hand over the territory.[120] To protect its members' legal interests the well guild investigated the case and petitioned the court to grant the territory to Jiang.[121]

The well guild continued to represent its members in negotiations with the state. For example, water carrier Liu rented a bitter-water well from the Public Works Bureau and transformed it into a foreign well. When the bureau decided to take over the well and fill it in because it blocked traffic, the guild petitioned the bureau to extend the deadline and compensate Liu seven hundred *yuan* for the well-drilling he had undertaken.[122] To maintain cohesion, the well guild would sometimes use violence in dealing with undisciplined carriers. The owner Wang ran a water business with one carrier at Tianqiao. He discovered

117 *Beiping chenbao*, April 23, 1936. (北平晨報 1936 年 4 月 23 日 第六版 衛生局催辦水井及水夫登記。)

118 *Jingbao*, May 3, 1936. (京報 1936 年 5 月 3 日第六版 不加入同業公會者準由各公會擬定制裁方案。)

119 Beijing Municipal Archives, *Hengxingquan shuijing puzhang Xie Zhonglin konggao Wang Zhiqing deng qiangzhan shuidao*, J181-21-52078, January 24, 1936. (北京市檔案館：恒興泉水井鋪掌解鐘麟控告王誌清等強占水道，J181-21-52078，1936 年 1 月 24 日。)

120 Beijing Municipal Archives, *Beipingshi gonganju waisanqu jingshu guanyu Chen Yuling yu Liu Jinqi yin zhengshui huou yian*, J181-021-08877, April 3, 1930. (北京市檔案館：北平市公安局外三區警署關於陳玉伶與劉金啓因爭水互相毆一案，J181-021-08877，1930 年 4 月 3 日。)

121 Beijing Municipal Archives, *Beijing tebieshi gonganju waisanqu chahuo Jiang Peiyi deng wei zhengshuidao juzhong qunou yian*, J181-021-08881, April 9, 1930. (北京市檔案館：北京特別市公安局外三區查獲江培義等爲爭水道聚眾群毆一案，J181-021-08881，1930 年 4 月 9 日。)

122 Beijing Municipal Archives, *Beijingshi jingye tongyehui wei huiyuan Liu Shanxing han Beijingshi shanghui*, J71-1-406, December 19, 1937. (北京市檔案館：北京市井業同業會函北京市商會，J71-1-406，1937 年 12 月 19 日。)

that a carrier named Zhang sold water in secret on the same street. Wang and his carrier went to argue with Zhang, and the discussion escalated into a fight with shoulder poles.[123] Zhang was notorious for encroaching on others' territories, but he was a large man and his peers were afraid of him. One carrier, Huang, who opposed Zhang's usurping of others' territories, invited three dozen carriers from the well guild to join him in fighting Zhang. Zhang also tried to gather other carriers, but no one responded. Zhang thus had to confront dozens of carriers by himself; he was struck on the head with a rod and lost consciousness.[124]

Throughout the summer in 1936, although the Public Health Bureau consistently assured the well guild that the purpose of registration was to facilitate administration rather than to confiscate distribution territories, the hesitant carriers believed the Bureau intended to take over the distribution territories and transform them into governmentally operated areas.

The well guild constantly petitioned to repeal registration by arguing that the territories were complicated and numerous conflicts remained unresolved. The chairman of the well guild also met with the director of the Public Health Bureau. After arduous negotiations, the bureau decided that registration could be limited to only basic information about owners and carriers, could exclude distribution territories, and could be completed by the well guild itself.[125] In July, the bureau discussed specific methods of registration with the leaders of the well guild.[126]

The registration of water carriers went smoothly once the guild began registering its own members. Water distribution territories were not included in the registration.[127] By the end of September, registration was complete: a total of 325 drinking-water wells, 1,674 owners and carriers, and 1,184 water wheelbarrows were registered at the Public Health Bureau.[128] The Bureau issued licenses for carriers at five cents per person and plates for water wheelbarrows

123 *Jingbao*, January 1, 1935. (京報 1935 年 1 月 1 日第六版 昨日天橋富長街爭水道兇毆 。)

124 *Jingbao*, April 4, 1935. (京報 1935 年 4 月 4 日第六版 水夫互爭水道。)

125 *Huabei ribao*, July 10, 1936. (華北日報 1936 年 7 月 10日 第六版 衞生局變通水夫登記辦法。)

126 *Huabei ribao*, July 15, 1936. (華北日報 1936 年 7 月 15 日 第六版 衞生局召集水夫代表談話。)

127 *Huabei ribao*, August 30, 1936. (華北日報 1936 年 8 月 30 日 第六版 水夫登記進行順利。)

128 *Beiping chenbao*, October 2, 1936. (北平晨報 1936 年 10 月 2 日 第六版 本市井商水夫登記竣事。)

at ten cents each.[129] When the aggressive state presented a formidable threat to the water regime, the well guild played the role of mediator between the state and the water carriers, and it successfully dismissed the state's effort to register water distribution territories. The carriers' continued control of distribution territories was key to their abiding success and power.

4 Down with the Water Lords: the Residents and the Carriers

As the water regime emboldened the carriers and they developed a reputation for providing poor service and water of dubious quality, residents' misgivings about Beijing's water supply grew and they called on the authorities to investigate the water suppliers. Prompted by the outcry, the new state came under pressure to reform the water supply and to reorder the urban environment.

Despite hygienic campaigns and state interventions, the indispensable water carriers persisted in monopolizing the water supply. When the state introduced regulations that created extra operational costs, carriers would employ their longstanding tactic of increasing prices against the residents. Thus, the effects of this power dynamic constantly led to friction between carriers and residents.[130] Conflict mounted as new inhabitants arrived after the fall of Manchuria in 1931; its occupation by the Japanese caused numerous migrants from the northeast provinces to swarm into Beiping. As strangers and newly arrived inhabitants, they were susceptible to bullying and extortion by water carriers.

Water became comparatively expensive because of the relocation of the national capital, which impoverished the residents of Beiping after 1928. A survey of twelve hundred poor households indicated that their water bills accounted for 3.6 percent of their total expenses, while a reasonable percentage was considered 2 percent.[131] Accordingly, ongoing conflicts between residents and water carriers arose because of water prices. For example, water carrier Kang rented a government well on West Street. Chef Liu, who worked on the inspection team, bought two buckets of water per day with a monthly

129 *Beiping chenbao*, September 19, 1936. (北平晨報1936年9月19日第六版本市水夫登記已達二百余家。)

130 During the 1930s almost four-fifths of Beiping's population relied on water carriers' water. Even in 1946 piped households in Beiping accounted for only 23.4 percent of the total population of 1,241,983. See *Beiping shizheng tongji* (Beiping, 1947), 33. (北平市政府統計室編：北平市政統計公用資料專輯，1947年，33。)

131 Niu Naie, "Beiping Yiqian erbai pinhu zhi yanjiu," *Shehui xuejie* 7 (1933): 172. (牛鼐鄂：北平一千二百貧戶之研究，社會學界 ，1933年第7卷。)

service payment. However, Liu did not pay the bill and instead asked for more buckets of water each day, which caused a fight.[132] Carrier Zhang supplied water to Zhu's household with a monthly subscription. However, he suspended the water supply for two days because of unpaid water bills, which gave rise to a quarrel.[133] Late water delivery was another problem that led to daily friction. Water carrier Wang had an altercation with the Huang family because of delayed water delivery, with the entire family joining in a brawl with Wang.[134] The Yin household asked water carrier Shen to deliver two buckets of drinking water each day. However, Shen did not bring the water regularly. The Yin household found another water carrier who offered to deliver four buckets each day, which caused antagonism between carrier Shen and the Yin household.[135]

Mischievous boys in the neighborhood also intensified the relations between carriers and their customers. Water carrier Wang had a well at Desheng Gate and supplied water to the residents. Wang scolded the son of a night-soil collector, Guo, for playing near the well. Guo argued and then fought Wang with his night-soil ladle.[136] Water carrier Liu left his water wheelbarrow unattended in the street when distributing water to his customer's kitchen. A naughty boy unplugged the spigot of the water wheelbarrow and drained the water. Liu was so furious that he fought with the boy's parents.[137] Since Beiping's society was somewhat pacific in the early 1930s, with few fights occurring between different groups, these daily clashes further reinforced the militant image of the water lords and exacerbated the fragile relations between water carriers and society at large.

The residents attempted to draw on new approaches to wrestle with the water lords. Following the orders of the Nationalist government, in 1930 the community founded a self-governing committee to engage in public issues, of which water supply was one. The locally elected assemblies discussed solutions to the water supply situation and warned water carriers not to oppress

---

132 Beijing Municipal Archives, *Kang Jian jiapao yuren an*, J183-002-04631, September 7, 1932. (北京市檔案館：康吉安加暴行於人案，J183-002-04631，1932 年 9 月 7 日。)

133 Beijing Municipal Archives, *Zhu Boquan deng kongg Zhang Ligang shanghai yian*, J183-002-05818, July 8, 1933. (北京市檔案館：朱伯泉等控張立剛傷害一案，J183-002-05818，1933 年 7 月 8 日。)

134 *Jingbao*, March 27, 1933. (京報 1933 年 3 月 27 日 6 版 娘子軍勢不可擋 水夫身陷重圍。)

135 Beijing Municipal Archives, *Neisanqu biaosong Chen Xiangran konggao tiaoshuiren Shen Qingfu buzhao shuijia an*, J181-021-14824, April 18, 1932. (北京市檔案館：內三區表送陳想然控告挑水人沈慶福不找水價案，J181-021-14824，1932 年 4 月 18 日。)

136 *Huabei ribao*, September 13, 1935. (華北日報 1935 年 9 月 13 日 第六版 糞閥戰水閥。)

137 *Shijie ribao*, May 26, 1930. (世界日報 1930 年 5 月 26 日第六版 拔開水塞子 水流得好看。)

the residents.[138] A proposal by the vice-chairman of the self-governing committee, Zhu Qinghua, to drill a foreign well in each of the 317 self-governing streets was approved.[139] Since the cost of drilling one foreign well was three hundred *yuan*, the self-governing committee would subsidize one hundred *yuan* for each well, and the residents of each street would pay the remaining sum and own the well. Residents of three small and poor streets petitioned to pool their subsidies and drill one well together. Residents on another street petitioned not to drill a new well since they had a governmental foreign well with excellent water.[140] In the end, however, the self-governing committee failed to drill any new wells due to a shortage of funds.

During the municipal government's hygiene movement, residents who comprehended the idea of public health actively used hygiene concepts as the basis of their claims when pushing back against the water lords. At one well house the Shandong owner was fond of goldfish and raised a few foot-long fish in his water trough. As a result, fish food and excrement contaminated the water, especially in the summer. Along with that, the coverless water tank was exposed to rain and dust. The residents who drank this water suffered constantly from dysentery and diarrhea. One resident whose family was afflicted with these diseases asked the owner to clean and cover the water tank. However, the overbearing owner was so insolent and stubborn that he ignored this resident, which aroused the indignation of the neighborhood. Some even proposed secretly poisoning the goldfish, but the idea was dismissed because it would poison the residents as well.[141]

To resist the manipulation of water carriers, some residents would open their own wells and sell water to their neighbors. In response, the well guild would thwart these residents' challenges to the water distribution territories. For example, one resident in the western section of the city drilled a new well and hired one carrier to sell water, which encroached on water carrier Chong's distribution territory and was therefore shut down by the well guild.[142] Another example is the piped households with monthly subscriptions and unlimited water usage who could challenge water distribution territories by selling their running water. The well guild kept a close eye on this and discovered in 1937

138 *Shishi baihuabao*, August 4, 1930. (實事白話報 1930 年 8 月 4 日 水閥。)

139 *Huabei ribao*, October 19, 1930. (華北日報 1930 年 10 月 19 日 第六版 每一自治街鑿較深洋井一口。)

140 *Huabei ribao*, November 5, 1930. (華北日報 1930 年 11 月 5 日 第六版 各自治街鑿井計劃。)

141 *Beiping chenbao*, July 21, 1933. (北平晨報 1933 年 7 月 21 日 可惡的水閥 水槽養金魚。)

142 *Shijie ribao*, September 4, 1934. (世界日報 1934 年 9 月 4 日第八版 西城玉皇閣水閥二次激戰。)

that thirty-three piped households were selling running water privately to their neighbors. This was against the regulations of the Waterworks and also conflicted with the water carriers' business. The well guild petitioned the chamber of commerce to ask Capital Waterworks and the municipal government to forbid this illegal act on the part of the households.[143]

Within the neighborhoods, merchants continued to play a significant role in dealing with the water lords. The water source for one well, which earned the name "Honeypot," located near the moat outside Fucheng Gate (阜成門), was a well-known sweet artesian spring. In 1885, the government dredged a moat and planned to dismantle the well, but was stopped by the local merchants and gentry. The water from this well was so pure and clean that residents in the western section of the city drew water from it. Under Empress Dowager Cixi's order, the imperial household also occasionally dispatched people to fetch its water. It was said that Cixi gave it the name of "Honeypot." In April 1932, one carrier, Xu, rented the well, built a well house, and started to sell the water. This irritated the community, and someone set fire to the well house at night. The well house burned down and paralyzed the use of the well. In October 1932, after petitioning the authorities, the forty-seven merchants in the neighborhood held a great ceremony to reopen the well and decided to join together to rent it from the Public Works Bureau.[144] The merchants further petitioned to have their rental fees exempted because of the city's economic downturn in 1936 and were approved.[145]

The case of the Honeypot well demonstrates the power of the local community, but very few neighborhoods were fortunate enough to have a merchant community with such cohesion. The drinking water at Lama Temple showcases the residents' vulnerability when confronting water lords. Lama Temple hired Shang to carry water from the well at temple, but the nearby residents in Fangjia Hutong (方家胡同) also used this well, and conflicts about drawing water sometimes occurred between Shang and the residents. To address this problem, the temple asked Shang to sell water to the residents. However, the root problem was that Shang did not have rights to this water distribution territory. Lured by profits, he was bold enough to hire a carrier to deliver water,

---

143 *Beiping chenbao*, January 6, 1937. (北平晨報 1937 年 1 月 6 日 第五版 私售自來水 市商會請查禁。)

144 *Minzhongbao*, May 11, 1941. (民眾報 1941 年 5 月 11 日 第五版 阜成門外的蜜罐井。)

145 Beijing Municipal Archives, *Liu ziqing deng qing jiao fuwai shuijing zuzhao*, J017-001-01358, April 1936. (北京市檔案館：劉子清等請繳阜外水井租照，J017-001-01358，1936 年 4 月。)

which led to a direct conflict with Huang, whose water distribution territory surrounded Lama Temple.[146] The constant conflicts turned Fangjia Hutong into a battlefield. None of the carriers managed to distribute water there, and the merchant Liu, allied with thirty merchants and residents suffering from the resulting water shortage, petitioned the police to order Shang to resume supplying water.[147] This appeal did not bring results, and the fighting between the two carriers persisted, which reveals the extent to which residents suffered from the water lords.

5 Conclusion

To control the flow of water was to control the flow of power. During the Nanjing decade the state's water governance in Beiping sought to discipline the water regime by creating hygienic water and carriers during state building. However, the concept of hygienic modernity promoted by the municipal government was unable to alter the way that water was collected and delivered. State-building in the nationalist government period was aimed at dismantling the power of water carriers, but their campaigns barely addressed the issue of water distribution territories, which were the cornerstone of the water regime. The elimination of the water carriers in favor of public control thus became unrealistic; on the contrary, water carriers under the patronage of the well guild successfully defended their territories and transferred the pressures brought on by the hygienic campaigns to residents, who consistently suffered from the water lords.

Preserving the distribution system was key to concentrating water power in the hands of carriers and accounted for the persistence of water carriers despite the increasing presence of the national government in Beiping. Although water carriers in Beijing were notorious for offering poor water quality and charging arbitrary fees, the municipal government had no ability to rebuild the water regime by reforming the water distribution territories. It was fully aware of this predicament, but a government-operated public service for

146 Beijing Municipal Archives, *Xuan Jixing bao sanyi pingmin shoushuichang gongyou bei Huang Qingshan ouda an*, J183-2-7470, August 22, 1933. (北京市檔案館：軒吉興報三義平民售水廠工友被黃慶山毆打案，J183-2-7470，1933 年 8 月 22 日。)

147 Beijing Municipal Archives, *Shangmin Liu Songquan deng chengqing suling Shang Xingwang yu gejia songshui*, J183-002-07218, June 30, 1933. (北京市檔案館：商民劉松泉等呈請速令商興旺與各家送水，J183-002-07218，1933 年 6 月 30 日。)

water was far from being implemented.[148] Meanwhile, the leftist intellectuals who supported the Communist Revolution redefined the labor and the meaning of the water lords by specifying that "water lords" meant well house owners but not ordinary water carriers. This proved to be an efficient approach under the communist's water governance for nationalizing water and deconstructing the water regime in the 1950s.

148 *Jingbao*, January 13, 1935. (京報 1935 年 1 月 13 日 第六版 取締水房二閥 水夫須註重清潔。); Huabei ribao, December 21, 1934. (華北日報 1934 年 12 月 21 日 第六版 平市三霸。)

CHAPTER 6

Epilogue: The Water Regime after 1937

In July 1937, the Japanese occupation of Beiping interrupted the Nationalists' water governance, but the Japanese puppet municipal government did not overrun the water lords. Only when the Communist party took over in 1949 would the CCP nationalization of the water system mark the end of the water carriers' regime and power. Moreover, while water nationalization itself is something that western states achieved in early twentieth century as part of their transition into modernity, water municipalization in Beijing does not present a similarly straightforward nationalization story. The water governance of the Japanese-occupation authority seems to have been consistent with the Nationalist government's governance; but the Communist Party completely reversed the pattern of government participation by directly involving itself in both water supply and distribution, and leaving no room for water carriers and their regime.

1 The Water Regime under Japanese Rule

After the fall of Manchuria in 1931, the Japanese steadily encroached on northern China, and on July 7, 1937, the Marco Polo Bridge Incident triggered the Second Sino-Japanese War (1937–1945), which is known in China as the War of Resistance and is considered part of World War II. On August 18, the Japanese captured and occupied Beiping with little resistance coming from inside the city. Thus, the city and its residents were spared the urban destruction that many other cities suffered later during the war.

During this wartime period the Japanese made Beiping the capital of its puppet state, renaming it Beijing. Soon thereafter, the Japanese commenced reconstruction of Beijing's water supply system because numerous Japanese migrant families had moved to the city and the running water was insufficient. The city's waterworks had used the Sun River as a water source since 1909, but the supply of water was constantly affected by shortages and poor quality. By 1939, 158,603 out of 185,395 households in Beijing relied on water from carriers, which was drawn from 395 wells.[1] In 1942 the Japanese decided to shift

1 Beijing tebieshi gongshu weishengju, ed., *Beijing tebieshi gongshu weishengju yewu baogao* (Beijing, 1941), 104. (北京特別市公署衛生局編：北京特別市公署衛生局業務報告, 1941, 104。)

 | DOI:10.1163/9789004720800_008

the source of running water from the river to underground water. In addition to the reconstruction of the running water system, the Japanese continued to carry out the hygienic campaigns that had been in effect since the Nanjing decade.

Under Japanese rule, the Beijing government put more emphasis on public health and water sanitation, and placed water and its carriers under rigid scrutiny. Water carriers and their wells were constantly targeted for sanitation inspection. In 1938, at least fifteen water carriers were fined because of an unclean water supply. For example, water carrier Li raised goldfish in his water tank and was fined thirty cents. Water carrier Zhao, who put his chamber pot near his well, was fined fifty cents. The owner of Shang's well was located close to a public toilet and he was ordered to demolish it.[2] In 1939 alone, wells were inspected and scrutinized 17,968 times by the Bureau of Public Health; the owners of unqualified wells were instructed to improve them, were admonished, and even fined.[3] In 1940 a cholera pandemic occurred in Beijing. To combat cholera's spread, the Bureau of Public Health ordered the collection and analysis of fecal samples from all the water carriers.[4]

Owing to the Japanese military dictatorship, the well guild restarted the exercise of self-discipline, which had been inconsistently practiced since the Nanjing decade. One tactic for surviving the state's intervention was the standard-sized bucket. In 1936, the well guild had agreed to a policy of using standard buckets to distribute water. These standard buckets were made and stamped with the seal of the well guild, but with the coming of war the water carriers had never fully implemented this policy. In April 1938, the Bureau of Public Health reissued the policy and strove to popularize the standard buckets.[5] However, the well guild immediately discovered that some carriers used fake standard buckets with the well guild's seal counterfeited by bucket craftsman Zhang.[6]

---

2 Beijing Municipal Archives, *Weishengju chahuo shoushui bujie an*, J183-002-27462, 1938. (北京市档案馆：卫生局查获售水不洁案, J183-002-27462, 1938 年。)

3 Beijing tebieshi gongshu weishengju, ed., *Beijing tebieshi gongshu weishengju yewu baogao* (Beijing, 1941), 105. (北京特別市公署衛生局編：北京特別市公署衛生局業務報告, 1941, 105.)

4 "Xunling Beijing jingye tongye gonghui," *Shizheng gongbao*, no. 96 (1940). (訓令北京井業同業公會 市政公報 1940 (96) 。)

5 "Tonggao jingshang ying zunzhao guiding shoushui jiage yizhao huayi," *Shizheng gongbao*, no. 17 (1938): 54. (通告井商應遵照規定售水價格以照劃一，市政公報 第十七期 通告 1938 (17) 第 54 頁。)

6 Beijing Municipal Archives, *Jingchaju waierqu Wang Ziyuan deng kong Zhang Qingxuan weizao jingye gonghui dayin yian*, J181-023-01266, May 14, 1938. (北京市檔案館：警察局外二區王子元等控張慶宣僞造井業公會大印一案, J181-023-01266, 1938 年 5 月 14 日。)

Thus, in December 1938 the well guild launched a campaign to inspect water carriers' standard buckets. For example, the well guild discovered that carrier Sun was distributing water in buckets that were smaller than the standard buckets; he was fined five *yuan* and his buckets were confiscated.[7] The standard bucket campaign reached its height in February 1939 when the well guild discovered at least nine cases of misconduct due to resistance to standardization. For example, water carrier Cai used smaller buckets, claiming that his teenage son was unable to carry the larger standard buckets.[8] Water carrier Ji consistently refused to switch to the standard buckets,[9] while water carrier Du secretly modified his standard buckets by decreasing their volume.[10] The well guild reported these cases to the police. In another instance, the well guild found three water carriers were still using regular buckets instead of standard buckets, and escorted these carriers with their buckets to the police.[11] As a result, the water carriers were fined and their buckets were sent to the well guild to be modified.[12] In addition to standard buckets, the well guild also set water prices. Those who did not follow the fixed price were fined. For example, well owner Wang constantly sold water at a higher price than the fixed price, and he was fined five *yuan* by the well guild.[13] Well owners Yu and Shi

7 Beijing Municipal Archives, *Waisiqu chengsong Sun Wenli shuitong weigai huoyin shuiliang buzu yian*, J181-023-04453, December 1938. (北京市檔案館：外四區呈送孫文禮水桶未蓋火印水量不足一案，J181-023-04453，1938 年 12 月。)

8 Beijing Municipal Archives, *Neierqu song maoquan shuijing puzhang Tian Shiren deng weizhang shoushui an*, J181-023-07334, February 1939. (北京市檔案館：內二區送龍茂泉水井鋪長田世仁等違章售水案，J181-023-07334，1939 年 2 月。)

9 Beijing Municipal Archives, *Waisanqu chengsong Liu Baotang kong Ji Qingfu weizhang shiyong jiushi shuitong an*, J181-023-07330, February 7, 1939. (北京市檔案館：外三區呈送劉寶堂控姬慶福違章使用舊式水桶案，J181-023-07330，1939 年 2 月 7 日。)

10 Beijing Municipal Archives, *Waisiqu chengsong Du Lianhuai deng weizhang shoushui an*, J181-023-07329, February 6, 1939. (北京市檔案館：外四區呈送杜連懷等違章售水案，J181-023-07329，1939 年 2 月 6 日。)

11 Beijing Municipal Archives, *Neisiqu chengsong Wang Jinping deng jun shiyong jiushi shuitong an*, J181-023-07333, February 9, 1939. (北京市檔案館：內四區呈送王金平等均使用舊式水桶案 ，J181-023-07333，1939 年 2 月 9 日。)

12 Beijing Municipal Archives, *Neisiqu song Liu Chengan deng bu zunzhang shiyong biaozhui shuitong shoushui an*, J181-023-07331, February 9, 1939; *Neisiqu song Zhu Wanshan bu zunzhang shiyong biaozhui shuitong an*, J181-023-07332, February 9, 1939. (北京市檔案館：內四區送劉成安等不遵章使用標準水桶售水案，J181-023-07331，1939 年 2 月 9 日；內四區送朱萬山不遵章使用標準水桶案，J181-023-07332，1939 年 2 月 9 日。)

13 Beijing Municipal Archives, *Waiwuqu song Wang Shuqi buan biaozhui jiage shoushui an*, J181-023-05878, February 1939. (北京市檔案館：外五區送王樹琦不按標準價格售水案 ，J181-023-05878，1939 年 2 月。)

increased their water prices, and as a result were fined two *yuan* each.[14] To better manage and supervise water carriers, the well guild also updated water carriers' identification cards and water-wheelbarrow license plates.[15]

In wartime Beijing the well guild not only restrained and disciplined itself, it also adopted new technology. In 1943 the well guild had 306 well owners.[16] By the end of the civil war in 1948, sixty-three owners had switched from the windlass method to electric pumps to draw water from wells.[17] Nevertheless, since both the Japanese and the Chinese state were preoccupied with the war and had no energy and incentive to reform the water regime and bring an end to the water carriers' power, the water carriers' hegemony survived the war and the chaos. The water supply was still dominated by water carriers and their regime. The real challenge would come following the establishment of the Chinese communist regime in 1949.

2 The Water Regime under Communist Rule

In February 1949 the Communist Army marched into Beijing and gained control of the city. Soon after, Beijing became the national capital of Red China. Because drinking water in the city had always been associated with legitimacy, its control became one of the principal means employed by the Chinese Communist Party (CCP) to build its own legitimacy. The notable writer Lao She wrote that poor people had no access to running water and had to rely on wells in Beijing. Under the Nationalist government, drinking water had been a privilege of elites.[18] In contrast, the Communists believed that access to modern running water should be expanded to the proletariat. In March 1949

---

14 Beijing Municipal Archives, *Neisanqu chengsong Yu Jingui deng taigao shuijia*, J181-023-21712, March 17, 1939. (北京市檔案館：內三區呈送於進桂等高擡水價，J181-023-21712，1939 年 3 月 17 日。)

15 "Tonggao jingshang yang yizhao fubiao jingwang jingye gonghui jiaofei goumling haopai you," *Shizheng gonbao*, no. 55 (1939). (通告井商仰依照附表徑往井業公會繳費購領號牌由，市政公報 第五十五期 文電 1939 年。)

16 Beijing Municipal Archives, *Beijingshi geye gonghui huiyuan jiashu ji ziben deng xiang diaochabiao*, J071-001-00106-043, 1943. (北京市檔案館：北京市各業公會會員家數及資本等項調查表，J071-001-00106-043，1943 年。)

17 Beijing Municipal Archives, *Beipingshi jingshuiye tongye gonghui suoshu huiyuan shiyong dianli yangjing dizhi xingming qingce*, J006-003-00445, January 1949. (北京市檔案館檔案：北平市井水業同業公會所屬會員使用電力洋井地址姓名清冊，J006-003-00445, 1949 年 1 月。)

18 Lao She, "Wo reai xin Beijing," *Bataiye* (Beijing: Beijing yanshan chubanshe, 1997), 344. (老舍：我熱愛新北京，八太爺，北京：北京燕山出版社 ，1997 年，第 344 頁。)

the Communists assumed control of the Waterworks[19] and nationalized it in November.[20]

To popularize running water in the city of Beijing, the Longxugou (龍須溝) District was selected as a demonstration area. Longxugou had been a slum since the late Qing period, harboring the urban poor, prostitutes, and gamblers. The Communists installed running-water faucets, which they called "democratic faucets," and governed them through residential committees. These faucets were new units built by the CCP in urban neighborhoods. In 1950, Lao She wrote a drama titled *Longxugou* to praise the achievements of the new Communist regime. In the play, it is Chairman Mao who brings running water to the urban poor. From 1950 to 1952, the government built more than 1,100 public standpipes in the streets, and the percentage of the population with access to running water reached the surprisingly high number of 85.2 percent.[21]

Because the CCP favored running water, the water carriers and their water were considered outdated. This logic was continuous with the logic of the Nationalist Party, but the difference between the CCP and the KMT was that the CCP distinguished water carriers from well owners. Although water carriers were perceived as water lords in the pre-Communist period, a number of residents sympathized with the water carriers rather than blaming them, while the pro-communists deemed them proletarian. One author stated that water carriers fully deserved the title of water lords because of their unchecked power, but he also admitted that carrying water was a strenuous occupation, and although most of the carriers were rough and illiterate fellows, they were faithful and trustworthy. The author's friend Liang, the former chair of the well guild, was distinguished among carriers by his gentlemanly manner of speech and behavior.[22] Furthermore, as the author acknowledged, carrying water required intensive physical labor, and carriers even died from overexertion.[23] These deaths generated sympathy, while other water carriers won good reputations: Well-house owner Song returned the extra money a customer mistakenly paid.[24] When another water carrier, Liu, learned from the newspaper of

19 *Renmin ribao*, March 22, 1949. (人民日報 1949 年 03 月 22 日，第 2 版。)

20 *Renmin ribao*, November 25, 1949. (人民日报 1949 年 11 月 25 日，第 4 版。)

21 Beijing Municipal Archives, *Beijingshi chengshi gongshui de jiben qingkuang*, 001-006-02461, October 1956. (北京市城市供水的基本情况，001-006-02461，1956 年 10 月。)

22 *Shishi baihuabao*, August 4, 1930. (實事白話報 1930 年 8 月 4 日 水閥。)

23 *Beiping chenbao*, March 12, 1934. (北平晨報 1934 年 3 月 12 日 第六版 雨雪紛紛中一水夫暴卒。)

24 *Shijie ribao*, July 25, 1928. (世界日報 1928 年 7 月 25 日第七版 誠實的水車夫 臨財不苟。)

the drought and famine in Shanxi Province, he donated two *yuan* he had saved from his meager income to the relief fund.[25]

The arduous working conditions of the water carriers also invited public sympathy and understanding. One author portrayed the lives and habits of water carriers as follows: "In the early morning of winter, you would see many water carriers carrying heavy wheelbarrows in each hutong. The squeaking noise reminded people, huddling up in warm comfort, of the frigidity outside. Whether cloudy or windy, the water carriers pushed heavily loaded water wheelbarrows with all their strength. Their barrows were always stuck in the half-foot-deep mud of the hutong. Sometimes you would see a sweating father and son pushing a water wheelbarrow, resembling horses or mules struggling in the mud; was there any difference?"[26] The newspaper account not only evoked the strength of the water carriers, but also showed that their lifestyle was harsh. The author concluded that considering the significant role water carriers played in residents' lives, it was unfair to call them water lords.[27]

When it came to the proto-communist view about water carriers, as early as 1919, the leading figure in the May Fourth Movement and the cofounder of the Chinese Communist Party, Chen Duxiu (陳獨秀, 1879–1942), had categorized water carriers with other working class members as the proletariat, making them the mobilized subjects of communism.[28] Wu Boxiao (吳伯簫, 1906–1982), a Shandong native and leftist writer who was motivated by Marxism when studying at Beijing Normal University in 1925, joined a study group and propagated communism among the water carriers.[29] From the 1920s forward, this perception of water carriers as members of the working class was not unusual. Wang Tongzhao (王統照, 1897–1957), another leftist writer who was also a Shandong native, portrayed a water carrier's life in his novel and profoundly sympathized with water carriers as proletarian.[30] A newspaper went further by arguing that the title of water lord was only suitable for well-house owners, and that well owners exploited the ordinary water carriers as domiciled slaves. Water carriers worked ceaselessly, like horses or cattle, with little

25 *Shishi baihuabao*, July 6, 1930. (實事白話報 1930 年 7 月 6 日 社會新聞 推水夫熱心助陝災。)

26 *Shenbao*, September 13, 1935. (申報 1935 年 9 月 13 日 第 17，18 版 北平速寫（三）水夫。)

27 *Huabei ribao*, October 31, 1935. (華北日報 1935 年 10 月 31 日 第六版 水夫之生活與習慣。)

28 *Chenbao*, December 1, 1919. (陳獨秀，告北京勞動界，晨報，1919 年 12 月 1 日。)

29 Wu Boxiao, "Wuhuaguo-Wo he sanwen," *Wenxue pinglun*, no. 5 (1981): 70–75.（吳伯簫：無花果-我和散文，文學評論，1981 年第 5 期，70–75。)

30 *Chenbao fukan*, November 5, 1925. (王統照，水夫阿三，晨報副刊 1925 年 11 月 5 日。)

payment.[31] To a large extent, water carriers were exploited laborers who were members of the working class.

The Communist regime inherited this notion and claimed that water carriers were considered an exploited class, while well-house owners were regarded as an exploiting class and an impediment to running water and modernization. Thus, it was legitimate to reform the well owners and liberate the water carriers from exploitation. The CCP municipal government was finally able to focus on overseeing the supply and distribution of water by reforming the water carriers. In January 1950 the government completed its survey of the well guild, including its members, well locations, and capital. In order to dismantle the monopoly of the water carriers and their regime, the communist government began to register and transform water carriers into workers who would distribute water for the national waterworks. For example, owner Zhang had carried water in Beijing for forty years, and his well house was transformed into a running-water station in 1952.[32] By 1956 the government had nationalized all individual water businesses. Unlike western governments that municipalized urban water by purchasing private water companies, the CCP achieved municipalization by confiscating water businesses and annihilating water carriers. The municipal government became the only water supplier. There were hundreds of water lords in Nationalist Beijing, but only one water lord remained in Communist Beijing, the process which has striking similarities with the case of nightsoil industry in modern Beijing.[33]

Water in the city existed in a close relationship with political structures, economic systems, and thousands of human bodies that could not go for even a day without water. Then and now, Beijing receives its water by way of environmental, technical, and social intermediaries; all these factors form a regime, amid which and through which water supplies are collected and distributed

---

31 *Shenbao*, February 23, 1937. (申報 1937 年 2 月 23 日 第 19 版 水車夫。)

32 Beijing Municipal Archives, *Beijingshi renmin zhengfu gongshangju saying qiye sheli dengji shenqingshu*, 022-008-00089-004, 1952. (北京市档案馆：北京市人民政府工商局私营企业设立登记申请书，022-008-00089-004，1952 年。)

33 Like water carriers, night soil carriers consisted of Shandong migrants who had a veritable monopoly on the profession and maintained customary rights to ply a particular "shit route." The night soil carriers also organized a guild in reaction to bureaucratic efforts to manage them and violently resisted external regulatory interference; they were variously referred to as "shit lords." During the Republican era, Beijing's municipal government proved just as unsuccessful in regulating night soil carriers as it did in wresting control of the water provisioning system from water carriers. The nightsoil industry was finally nationalized by the communist regime after 1949. See Joshua Goldstein, *Remains of the Everyday: A Century of Recycling in Beijing* (Berkeley: University of California Press, 2021).

to householders. Since the Qing period, water carriers had rearranged nature through wells and through the apparatus of the well house, while water carts were sent through the neighborhoods to ensure water's regular distribution according to a complicated and delicate system in which the power of the water carriers was rooted in their territories and native-place ties. Their power remained potent and resilient through the arrival of Japanese well-drilling technology in the first decade of the twentieth century, the challenge of running water in 1910, and the introduction of state hygiene mandates in 1927.

The city's water supply, a key element in the city's metabolism, had major effects on the urban environment and its politics. The water regime constructed by the Qing state was employed as a means of social integration in the Inner City and a means of segregation in the Outer City, creating a spatial and social dichotomy that persisted into the late Qing period. Within their territories and as a power base in the back alleys of the city, water carriers remained a crucial force in defining the social fabric and domestic rhythms of the neighborhoods. These neighborhoods were fragmented into numerous water territories and residents were subjected to the regime of the water carriers who, as one of the largest non-state influences in the city's daily life, kept a tight rein on drinking water. By coopting well-drilling technology, manipulating political movements, and reorganizing their well guild, these so-called traditional water carriers deftly adapted to the dramatic changes brought about by modernization. The rise of the modern water enterprise, state-building, and the changing nature of neighborhoods prompted the transformation of water carriers into water lords, while individual residents ended up shouldering the unexpected burden of modernization because water carriers managed to internalize the pressure of competition generated by modern running water and transfer it to the city's inhabitants.

Although the Nationalist government sought to subdue water and its regime through hygienic modernization, it never took over the water carriers' service or reformed the water distribution territories. Meanwhile, the Waterworks failed to meet the challenge of urbanization and was unable to provide a plentiful, pure, and constant water supply. Instead, the water carriers managed to retain power throughout Beijing's transformations under various forms of government and political rule. Ultimately, however, by nationalizing the water business, the CCP's water governance reversed the fortunes of the water carriers and brought an end to non-statist approaches to water rights. Water carriers and their provisioning system, albeit highly resistant to outside management, performed an indispensable service by supplying the water resources that underfunded and infrastructurally inadequate modern utilities like Beijing's Capital Waterworks simply could not provide to most of the semiarid city's large population. The lack of viable alternatives meant that without the "water lords" the city would have gone thirsty. The history of water in Beijing belies the idea that the development of urban water systems follows a linear, progressive trajectory. Instead, Beijing's water system offers us an alternative modernity in which urban governance is characterized by accommodation and compromise between old and new.

Bibliography

Primary Sources

Archives

Archives of the Grand Secretariat, Institute of History and Philology, Academia Sinica, Taipei [中央研究院歷史語言研究所 內閣大庫檔案]

Beijing Municipal Archives [北京市檔案館]

Beiyang Diplomatic Archives at Institute of Modern History, Academia Sinica, Taipei [臺灣中央研究院近代史所檔案館]

Chinese First Historical Archives [中國第一歷史檔案館]

Chinese Second Historical Archives [中國第二歷史檔案館]

Niida Collection, Oriental Library, Tokyo University Library

Rockefeller Archives Center, New York City

Images

Princeton-in-Peking archives, Princeton University Library

Sidney D. Gamble Photographs Collection, Special Collection, Duke University

White Brother Collection, Division of East Asia, Library of Congress

Wilder Haydn Haines Papers, Princeton University Library

Newspapers

Aiguo baihuabao 愛國白話報 [Patriotic Mandarin News], 1914–1921

*Baihua jiebao*白話捷報 [Mandarin Express], 1913

Beijing baihuabao 北京白話報 [Beijing Mandarin News], 1920–1922

Beijing yishibao 北京益世報 [World Welfare News], 1917–1948

Beiping chenbao

Chenbao fukan

Chenzhong bao 晨鐘報 [Morning Bell News], 1917–1918

Dagongbao 大公報 [Ta Kung Pao], 1904

Dalubao 大陸報 [Mainland News], 1905

Huabei ribao 華北日報 [North China Daily], 1929–1938

Jingbao 京報 [Capital Post], 1928–1936

Jingdu shuigui

Jinghua ribao 京華日報 [Capital Daily], 1904–1923

Minzhongbao 民眾報 [Mass News], 1941

North China Herald and *North China Daily News*, 1850–1951

Peking Mandarin Daily [京話日報], 1904–1906)

Qianshuo huabao

*Renmin ribao*人民日報 [People's Daily], 1949–1950

Shenbao 申報 [Shanghai News], 1872–1949
Shijie ribao 世界日報 [World News], 1925–1949
Shishi baihuabao 實事白話報 [Factual Mandarin News], 1922–1930
Shuntian shibao 順天時報 [Shuntien Times], 1905–1930
South China Morning Post
Weisheng baihuabao 衛生白話報 [Hygiene Mandarin News], 1908
*Yishi baihuabao*益世白話報 [World Welfare Mandarin News], 1917

Journals

Beijing dangan shiliao 北京檔案史料 [Beijing Archival Materials], 1986–
Beijing dier weishengju shiwusuo nianbao 北平市衛生處第二衛生區年報 [Annual report of the First Health Station], 1934–1936
Beijing diyi weishengju shiwusuo nianbao 北平市衛生處第一衛生區年報 [Annual report of the First Health Station], 1926–1937
Beiping shizheng gongbao 北平市政公報 [Beiping Municipal Bulletin], 1929–1937
The Japan Weekly Mail, 1902
The Missionary Herald, 1913
Qingdai dangan shiliao congbian 清代檔案史料叢編 [Compilation of archival materials in Qing], 1978–
Weisheng yuekan 衛生月刊 [Hygienic Monthly]

Oral History

Grant, John Black, *The Reminiscences of Doctor John B. Grant*, Columbia University Oral History Collection; part. 3, no. 195

Secondary Sources

Allied Powers, Supreme Commander. *Preliminary Study*. Natural Resources Section, issues 49–64 (1951).

Archer, M. T. "Drilling in Tzu-Liu-Ching, China." In *History of Petroleum Engineering,* 146–52. Dallas, TX: American Petroleum Institute, 1961.

Asen, Daniel. *Death in Beijing: Murder and Forensic Science in Republican China.* Cambridge: Cambridge University Press, 2016.

Bankoff, Greg, Uwe Lübken, and Jordan Sand, eds. *Flammable Cities: Urban Conflagration and the Making of the Modern World.* Madison: University of Wisconsin Press, 2012.

Baqi tongzhi chuji, vol. 33, 438–439. (八旗通誌初集 卷 23 營建誌一，吉林：東北師範大學出版社，1985 年，第 438–439 頁。)

Barbour, George B. "Water Supply of Tingchow and Shuntehfu Area." *Journal of the Association of Chinese and American Engineers* 3, no. 9 (October 1922): 8ff.

Barbour, George B. "Deep Wells in the Peking Area." *Bulletin of the Geological Society of China* 3, no. 2 (1924): 136.

Barbour, George B. "Artesian Well Supply at Haitieu." *Journal of the Association of Chinese and American Engineers* 2, no. 1 (1924): 18ff.

Barbour, George B. "Deep Wells in the Peking Area." *Bulletin of the Geological Society of China* 3, no. 2 (1924): 127–38.

Barbour, George B. "Further Data Regarding Deep Wells in the Peking Area." *Bulletin of the Geological Society of China* 9, no. 1 (1930): 49–58.

Barbour, George B. *In China When….* Cincinnati, OH: University Publications, University of Cincinnati, 1975.

"Beijing jingye diaocha" 北京井業調查 [Survey of well-drilling in Beijing]. *Xinhua* 興華 [Prosper China] 23, no. 32 (1926): 24–29.

Beijing Municipal Archives, ed. *Beijing dang'an shiliao* 北京檔案史料 [Beijing historical archive materials], 1986–.

Beijing Municipal Archives, ed. *Beijing zilaishui gongsi dangan* 北京自來水公司檔案 [*The Archives of Beijing Waterworks*]. Beijing: Beijing yanshan chubanshe, 1986.

Beijing niujie zhishu-Gangzhi. Beijing: Beijing chubanshe, 1990.

Beijing tebieshi gongshu weishengju, ed. *Beijing tebieshi gongshu weishengju yewu baogao*. Beijing, 1941).

Beijing tushuguan cang zhongguo lidai shike tuoben huibian 北京圖書舘藏中國歷代石刻拓本滙編. Zhengzhou: Zhongzhou guji chubanshe, 1990.

Beijingshi danganguan, ed. "Waicheng xunjing zongting wei yanna pohuai shuizhi zhitu dengshi zhi zilaishui gongsi yiwen" [外城巡警总厅为严拿破坏水质之徒等事致自来水公司移文]. In *Beijing zilaishui gongsi dangan shiliao*, August 19, 1910.

Beijingshi danganguan, ed. "Zilaishui gong ni zeyao luxu kuochong jihuashu." In *Beijing zilaishui gongsi dangan shiliao.* Beijing: Beijing yanshan chubanshe, 1986.

Beiping shi shizheng gongbao 北平市市政公報 [Official bulletin of the Beiping municipal government]. Beiping, 1928–1937.

Beiping shizheng tongji. Beiping, 1947.

Beiping shizhengfu canshi shi, ed. *Beiping shi shizheng fagui huibian* 北平市市政法規滙編 [A collection of administrative laws and regulations in Beijing]. Beiping, 1934.

Beiping shizhengfu weishengchu yewu baogao 北平市政府衛生處業務報告 [Work reports of the Beiping municipal department of public health]. Beiping, 1934.

Beiping tebieshi diyi weisheng shiwusuo disinian nianbao. Beiping, 1929.

"Beiping tebieshi weishengju 1929 niandu shizheng dagang" 北平特別市衛生局 1929 年度施政大綱 [The administrative plans of Beiping Municipal Department of Public Health in 1929]. *Beijing dang'an shiliao* 27 (1992).

Beipingshi dier weisheng shiwusuo dier niandu nianbao. Beiping, 1935.

Beipingshi weishengchu dier weishengqu shiwusuo diyi niandu nianbao. Beijing, 1934.

Bennett, Vivienne. *The Politics of Water: Urban Protest, Gender and Power in Monterrey, Mexico.* Pittsburgh, PA: University of Pittsburgh Press, 1995.

Blake, Nelson M. *Water for the Cities: A History of the Urban Water Supply Problem in the United States.* Syracuse, NY: Syracuse University Press, 1956.

Bredon, Juliet. *Peking: A Historical and Intimate Description of Its Chief Places of Interest.* Shanghai: Kelly & Walsh, 1920.

Breton, Joseph. *China: Its Costume, Arts, Manufactures &c,* vol. II. London: J. J. Stockdale, 1813.

Broich, John. *London: Water and the Making of the Modern City.* Pittsburgh, PA: University of Pittsburgh Press, 2013.

Bullock, Mary Brown. *An American Transplant: The Rockefeller Foundation and Peking Union Medical College.* Berkeley: University of California Press, 1980.

Chai, E. "Gengxin jishi." In *Yihetuan,* edited by Jian Bozan. Shanghai: Shanghai renmin chubanshe, 1957.

Chan, Hok-Lam. *Legends of the Building of Old Peking.* Hong Kong: The Chinese University of Hong Kong, 2008.

Chan, Wellington K. K. *Merchants, Mandarins, and Modern Enterprise in Late Ch'ing.* Cambridge, MA: Harvard University Asia Center, 1977.

Chang, Michael G. *A Court on Horseback: Imperial Touring & the Construction of Qing Rule, 1680–1785.* Cambridge, MA: Harvard University Press, 2007.

Chen, Janet Y. *Guilty of Indigence: The Urban Poor in China, 1900–1953.* Princeton, NJ: Princeton University Press, 2012.

Cheng, Weikun. *City of Working Women: Life, Space and Social Control in Early Twentieth-Century Beijing.* Berkeley: University of California Press, 2011.

Chi, Zehui 池澤匯, ed. *Beijing gongshangye diaocha* 北京工商業調查 (*A survey of industry and commerce of Beijing*). Beijing: Beijing shehui ju, 1932.

Chi, Zehui 池澤匯, ed. *Beipingshi gongshangye gaikuang.* Beiping: Beiping Social Bureau, 1932.

Chiu, C.L. 邱仲麟 "Shuiwozi: Beijing de minsheng yongshui" 水窩子: 北京的民生用水 [*Water den: Water carriers and daily water, 1637–1937*]. In *Zhongguo chengshi shenghuo* [*Life in urban China*], edited by H. T. Li, 203–52. Beijing, 2006.

Chou, Tung Pi 周同璧. "'The Bitter Water' and 'Sweet Water' in Peking." Undergraduate paper, Yenching University, 1926.

Chu, Weikai. "Yanjing zayong." In *Beijing fengsu zayong,* edited by Lei Mengshui. Beijing: Beijing Guji chubanshe, 1982.

Cohen, Paul A. *History in Three Keys: The Boxers as Event, Experience and Myth.* New York: Columbia University Press, 1997.

"Comments on the Deep Wells in the First Inner Health District." *Eighth Annual Report of the Department of Public Health,* June 30, 1933, Appendix II, p. xiv, RG 5.3, Series 601J, Box 220, F2740-2741, Rockefeller Archive Center, New York.

Crawford, Wallace. "The Salt Industry of Tzeliutsing." *China Journal of Science and Art* (Shanghai) 4 (1926): 174.

De, Shuoting. "*Caozhu yichuan.*" In *Qingdai Beijing zhuzhici,* edited by Lu Gong. Beijing: Beijing chubanshe, 1962.

Dinçkal, Noyan. "Reluctant Modernization: The Cultural Dynamics of Water Supply in Istanbul, 1885–1950." *Technology and Culture* 49, no. 3 (July 2008): 675–700.

Dodgen, Randall. *Controlling the Dragon: Confucian Engineers and the Yellow River in Late Imperial China.* Honolulu: University of Hawai'i Press, 2001.

Dong, Madeleine Yue. *Republican Beijing: The City and Its Histories.* Berkeley: University of California, 2003.

Dong, Madeleine Yue, and Joshua L. Goldstein, eds. *Everyday Modernity in China.* Seattle: University of Washington Press, 2006.

Dray-Novey, Alison. "Policing Imperial Peking: The Ch'ing Gendarmerie, 1650–1850." Ph.D. diss., Harvard University, 1981.

Dray-Novey, Alison. "Spatial Order and Police in Imperial Beijing." *The Journal of Asian Studies* 52, no. 4 (Nov. 1993): 885–922.

Dray-Novey, Alison. "The twilight of the Beijing Gendarmerie, 1900–1924." *Modern China: An International Quarterly of History and Social Science* 33, no. 3 (2007): 349–76.

Duara, Prasenjit. *Culture, Power, and the State: Rural North China, 1900–1942.* Stanford, CA: Stanford University Press, 1988.

Dudgeon, John. *Diet, Dress and Dwellings of Chinese in Relation to Health.* London: William Clowes and Sons, Limited, 1876.

Einhorn, Robin L. *Property Rules: Political Economy in Chicago, 1833–1872.* Chicago, University of Chicago Press, 1991.

Elliott, Mark C. *The Manchu Way: The Eight Banners and Ethnic Identity in Late Imperial China.* Stanford, CA: Stanford University Press, 2001.

Eyferth, Jacob. *Eating Rice from Bamboo Roots: The Social History of a Community of Handicraft Papermakers in Rural Sichuan, 1920–2000.* Cambridge, MA: Harvard University Asia Center, 2009.

Fang, Yiji 方頤積. "Beijingshi zhi jingshui" 北平市之井水 [Well water in Beiping]. *Weisheng yuekan* 衛生月刊 [Hygienic Monthly] (March 1928): 7–10.

Feuerwerker, Albert. *Industrial Enterprise in Twentieth-Century China.* Berkeley: University of California Press, 1967.

Fromann, C. W. *Die Bohr-Methode der Chinesen oder das Seilbohren*. Coblenz, 1835.

Fu, Jia. "Qinggong huiyi." In *Wanqing gongting shenghuo jianwen.* Beijing: Wenshi ziliao chubanshe, 1982.

Fu, Nongshi. "Dumen jibian baiyong." In *Qingdai Beijing zhuzhici,* edited by Lu Gong. Beijing: Beijing chubanshe, 1962.

Gamble, Sidney D. *Peking: A Social Survey.* New York: George. H. Doran Company, 1921.

Gandy, Matthew. *Concrete and Clay: Reworking Nature in New York City*. Cambridge, MA: MIT Press, 2002.

Gao, Yan. *Yangzi Waters: Transforming the Water Regime of the Jianghan Plain in Late Imperial China*. Brill, 2022.

"Gaunli yinshuijing guize." *Neizheng gongbao* 1, no. 3 (1928): 39–40. (管理飲水井規則部令第三九七號公布國民政府內政部內政公報 法規 1928, 1(3), 39–40。)

Gerth, Karl. *China Made: Consumer Culture and Creation of the Nation*. Cambridge, MA: Harvard University Asia Center, 2003.

Goldstein, Joshua. *Remains of the Everyday: A Century of Recycling in Beijing*. Berkeley: University of California Press, 2021.

Goossaert, Vincent. *The Taoists of Peking, 1800–1949: A Social History of Urban Clerics*. Cambridge, MA: Harvard University Asia Center, 2007.

Gordon, Charles Alexander. *Remarks on Certain Points Relating to Public Health in China*, vol. 19. London: William Clowes and Sons, Limited, China.

Gordon, C. A. *An Epitome of the Reports of the Medical Offices to the Chinese Imperial Maritime Customs Service, from 1871 to 1882*. London, 1885.

Goubert, Jean-Pierre. *The Conquest of Water: The Advent of Health in the Industrial Age*. Princeton, NJ: Princeton University Press, 1989.

Graham, Stephen, and Nigel Thrift. "Out of Order: Understanding Repair and Maintenance." *Theory, Culture & Society* 24, no. 3 (2007): 1–25.

Grant, John B. *Report on a General Health Survey of Peking, China*, February 1922, RF RG 5 Series 2, SS- 601, Box 55, F346, Rockefeller Archive Center, New York.

Guangxu Qingding daqing huidian zeli, vol. 1039.

Gugong bowuyuan, ed. *Qingding neiwufu zeli erzhong* (欽定內務府則例二种). Haikou: Hainan chubanshe, 2000.

"Gugong bowuyuan tushuguan zhanggubu" *Zhanggu congbian* (掌故叢編) (1928).

Guochao xianzhenglu, vol. 20. (國朝獻徵錄·卷 20 黄谏。)

Guo, Songyi. "Qingdai Beijing de Shandong yimin." *Zhongguo shi yanjiu*, no. 2 (2010): 27–46. (郭松义：《清代北京的山东移民》，《中国史研究》，2010 年第 2 期，第 27–46 页。)

Han, Guanghui 韓光輝. *Beijing lishi renkou dili* 北京歷史人口地理 [Historical Geography of the population in Beijing]. Beijing: Beijing daxue chubanshe, 1996.

Harrell, Paula. *Sowing the Seeds of Change: Chinese Students, Japanese Teachers, 1895–1905*. Stanford, CA: Stanford University Press, 1992.

Harrison, Henrietta. *The Making of the Republican Citizen: Political Ceremonies and Symbols in China, 1911–1929*. New York: Oxford University Press, 2000.

Hart, John W. "Shanghai Water-Works." *Minutes of the Proceedings*, ICE, 100, January 1890.

Hassan, John. *A History of Water in Modern England and Wales*. Manchester: Manchester University Press, 1998.

Hattori, Unokichi服部宇之吉, et al. *Pekin shi* [*Beijing Gazetteer of Late Qing Period*]. Tokyo: Hakubunkan, 1908; translated by Zhang Zongping, Qing mo Beijing zhi ziliao 清末北京志資料. Beijing: Beijing yanshan chubanshe, 1994.

He, Gangde. *Huamengji,* vol. 1 (何剛德：活夢集 卷上。)

Heyking, Elisabeth von. *Memoirs of Four Continents, 1886–1900,* cited in Felicitas Titus, *Old Beijing: Postcards from the Imperial City.* Tokyo: Tuttle Publishing, 2012.

Hibbert, Eloise Talcott. *Jesuit adventure in China: During the reign of K'ang His.* New York: E. P. Dutton and Company, 1941.

Himley, Matthew. "Geographies of Environmental Governance: The Nexus of Nature and Neoliberalism." *Geography Compass* 2 (2008): 433–51.

Hooker, Mary. *Behind the Scenes in Peking, Being Experiences during the Siege of the Legations.* London: John Murray, 1911.

Hou, Renzhi 侯仁之. *Beijing chengshi lishidili* 北京城市歷史地理 (*The Historical Geography of Beijing*). Beijing: Beijing yanshan chubanshe, 2000.

Hu, Zhang 胡璋. *Dongguo zaojing fa,* 東國鑿井法 [*The method of Japanese well-drilling*]. Published by Agricultural Society, Guangxu Period (1874–1908), rare book housed at Nanjing Library, China. (農學會刊行，光緒年間，南京圖書館古籍部藏。)

Huang, Wanjie. "Beiping yinshuijing zhi wuran laiyuan yuqi gaishan fangce." *Xinyiyao* 3, no. 9 (1935): 767–71. (黃萬傑：北平飲水井之汙染來源與其改善方策，新醫藥 1935 年第 3 卷 9 期, 767–71。)

Hunt, Michael. "The Forgotten Occupation: Peking, 1900–1901." *Pacific Historical Review* 48 (November 1979): 501–29.

Iijima, Wataru. "Jindai Beijing gonggong weisheng zhi kaocha-20 shijie chuqi de zhongguo guojia weisheng." In Beijingshi danganguan bian, *Dangan yu beijingshi guoji xueshu taolunhui lunwen.* Beijing: Zhongguo dangan chubanshe, 2003, 68–74. (飯島涉：《近代北京公共衛生之考察—20世紀初期的中國 "國家衛生" 》, 北京市檔案館編：《檔案與北京史國際學術討論會論文》（下冊）北京：中國檔案出版社， 2003 年，第 68–74 页。)

Im, Ki-jung 林基中, ed. *Yŏnhaengnok chŏnjip* 燕行錄全集 [A collection of travelogues written by Korean envoys who traveled to China between the 13th and 19th centuries], 100 vols. Seoul: Tongguk Taehakkyo Chulpanbu, 2001.

Ji, Yun 紀昀. *Yuewei caotang biji* [閱微草堂筆記]. Shanghai: Shanghai guji chubanshe, 1980.

Jian, Bozan, ed. *Yihetuan.* Shang: Shanghai renmin chubanshe, 1957. (翦伯贊等編 義和團 上海：上海人民出版社，1957 年。)

Jie. "Jingshui xiaodu de yiyi." *Beiping yikan* 2, no. 9 (1934): 35–36. (傑：井水消毒的意義 北平醫刊 1934 年 10 月 第 2 卷第 9 期，第 35–36 页。)

Jin, Shoushen 金受申. *Lao Beijing de shenghuo* 老北京的生活 [Life in old Beijing]. Beijing: Beijing chubanshe, 1989.

Jingdu shizheng gongsuo 京都市政公所, ed. *Jingdu shizheng huilan* 京都市政滙覽 [A collection of materials about municipal administration of the capital metropolis]. Beijing: Jinghua shuju, 1919.

"Jingshi jingchating gaojie zhuyi shuijing weisheng." *Shizheng jikan*, no. 1 (October 1925): 27. (市政季刊 1925 年 10 月 第 1 期 京師警察廳告誡註意水井清理，第 27 頁。)

Jobard, J. B. "Uber das Brunnenbohren mit dem Seil." *Polytechisches Journal* (Stuttgart) 105 (1847): 14–24.

Kim, Ch'ang-ŏp 金昌業. "*Yŏnhaeng ilgi*" 燕行日記 [Diary of a China trip]. In *Yŏnhaengnok chŏnjip* 燕行錄全集 [A collection of travelogues written by Korean envoys who traveled to China between the 13th and 19th centuries], 100 vols., edited by Im Ki-jung 林基中. Seoul: Tongguk Taehakkyo Chulpanbu, 2001.

Koeppel, Gerard T. *Water for Gotham: A History*. Princeton, NJ: Princeton University Press, 2000.

Lai, Huimin 賴慧敏. *Danwen qimin: Qingdai de fanlv yu shehui* 但問旗民. Taipei: Wunan tushu chuban gongsi, 2007.

Lao, She. "Wo reai xin Beijing." *Bataiye*. Beijing: Beijing yanshan chubanshe, 1997, 344. (老舍：我熱愛新北京，八太爺，北京：北京燕山出版社 ，1997 年，第 344 頁。)

Lao, She. *Zheng hongqi xia* 正紅旂下 [The red banner]. Beijing: Renming wenxue chubanshe, 1980.

Lei, Mengshui, ed. *Beijing fengsu zayong*. Beijing: Beijing Guji chubanshe, 1982. (雷夢水，北京風俗雜詠 北京：古籍出版社，1982 年。)

Lemos, Maria Carmen, and Arun Agrawal. "Environmental Governance." *Annual Review of Environmental Resources* 31 (2006): 297–325.

Li, Chunfeng. *Commentaries and Illustrations of Well-Drilling*. Masters, Continuation of Imperial Collection of Four, 599–608.

Li, Jung. "An Account of the Salt Industry at Tzu-liu-ching," translated by Lien-che Tu Fang. *Isis* 39, no. 4 (1948): 228–34.

Li, Lilian M., and Alison Dray-Novey. "Guarding Beijing's Food Security in the Qing Dynasty: State, Market and Police." *Journal of Asian Studies* 58, no. 4 (November 1999): 992–1032.

Li, Lillian, Alison Dray-Novey, and Haili Kong. *Beijing: From Imperial Capital to Olympic City*. New York: Macmillan, 2008.

Li, Wan. *Jingcheng guji kao* (勵萬，京城古跡考).

Li, Youying. "Survey of Laborers in Beijing." *New Youth* 7, no. 6 (1920): 14–17. (李幽影：《北京劳动状况》，《新青年》第 7 期第 6 号，1920 年 5 月，第 14–17 页。).

Lin, Yutang. *Moment in Peking*. New York: John Day Co., 1939.

Liu, Chunyuan. "Woguo songdai jingyan zuanzhao gongyi de zhongyao gexin-Sichuan zhuotongjing." *Wenwu* 23 (December 1977): 66–72.

Liu, Ruoyu. *Zhuo zhonghu*. (劉若愚：《酌中誌》。)

Liu, Xiaomeng 劉曉萌. *Qingdai Beijing qiren shehui* 清代北京旗人社會. Beijing: Zhongguo shehui kexue chubanshe, 2008.

Longwang pusa lingjing ji. *Beijing tushuguan cang zhongguo lidai shike taben huibian.* Zhengzhou: Zhongzhou guji, 1772.

Lu, Gong 路工, ed. *Qingdai Beijing zhuzhici* 清代北京竹枝詞 [The bamboo trigs in Qing Beijing]. Beijing: Beijing chubanshe, 1962.

Lu, Hanchao. *Beyond the Neon Lights: Everyday Shanghai in the Early Twentieth Century.* Berkeley: University of California Press, 1999.

Lü, Xun. *Lü Xun Riji.* Beijing: Renmin wenxue chubanshe, 1976. (《魯迅日記》上卷 北京：人民文學出版社，1976 年。)

Lu, Yi 陸毅. *Xuncheng suoji* 巡城瑣記 [The trivial records of city patrolling]. Department of Rare Books, Nanjing Library, China.

Ma, Zhao. *Runaway Wives, Urban Crimes and Survival Tactics in Wartime Beijing, 1937–1949*. Cambridge, MA: Harvard University Press, 2015.

Malott, Matthew Keith. *Nomine Caesaris: An Examination of the Propagandistic Functions of the Aqueducts of Rome during the Early Empire.* Classical Association of Canada Essays, 2004.

Mateer, A. H. *Siege Days: Personal Experiences of American Women and Children during the Peking Siege.* New York: Fleming H. Revell Company, 1903.

Millard, R. "Privatisation in Historical Perspective: The UK Water Industry." In *Money, Trade and Payments: Essays in Honour of Dr. Coppock,* edited by D. Cobham, 197. Manchester: Manchester University Press, 1989.

Mukhopadhyay, Asok Kumar. *Politics of Water Supply: The Case of Victorian London.* Calcutta: World Press, 1981.

Muscolino, Micah S. *The Ecology of War in China: Henan Province, the Yellow River and Beyond, 1938–1950.* Cambridge: Cambridge University Press, 2015.

Naquin, Susan. *Peking: Temples and City Life, 1400–1900*. Berkeley: University of California Press, 2000.

Needham, Joseph. *Science and Civilization in China.* Cambridge: Cambridge University Press, 1954.

"1934nian Beipingshi zhengdun gongce shiliao." *Beijing dangan shiliao*, no. 3 (2007): 5–10. (1934年北平市整頓公廁史料，北京檔案史料 2007 年第 3 期, 5–10。)

"1934nian Beipingshi zhengdun gongce shiliao." *Beijing dangan shiliao*, no. 3 (2007): 24–29. (1934年北平市整頓公廁史料，北京檔案史料 2007 年第 3 期, 24–29。)

Niu, Naie 牛鼐鍔. "Beiping yiqian erbai pinhu zhi yanjiu" 北平一千二百貧戶之研究 [A study of 1200 impoverished households in Beiping]. *Shehui xuejie* 社會學界 [*Sociological World*] 7 (June 1933): 147–87.

Niujie zhi 牛街志 (Gazetteer of Niujie). Beijing: Beijing chubanshe, 1990.

"Nonggongshangbu wei zilaishui gongsi kuochong zhuanguan bing guandu shangban xiaoshi zhi minzhengbu ziwen," (August 9, 1911). *Lishi dangan*, no. 2 (1992): 64.

Norman, F. J. “Boring in Japan.” *Transactions of the Institution of Mining Engineers* 23 (1903): 685–99.

“Obituary: John Dudgeon, M.D.” *The British Medical Journals* 1, no. 2098 (March 16, 1901): 679.

Ogle, Maureen. *All the Modern Conveniences: American Household Plumbing, 1840–1890*. Baltimore, MD: Johns Hopkins University Press, 1996.

Oriental Culture Institute of Tokyo University. *Dongyang wenhua yanjiusuo cang zhongguo tudi wenshu mulu*, 123–31. (濱下武誌等編，東洋文化研究所藏中國土地文書目錄·解說，東京：東京大學東洋文化研究所附屬東洋文獻中心，1986 年，第 123–131 頁。)

Pan, Jixing. “Zhongguo shenjing zuantan jishu de qiyuan, fazhan he xichuan.” *Yanyeshi yanjiu* 23 (October 2009): 3–33.

Perdue, Peter. *Exhausting the Earth: State and Peasant in Hunan, 1500–1850*. Cambridge, MA: Harvard University Press, 1987.

Pietz, David A. *The Yellow River: The Problem of Water in Modern China*. Cambridge, MA: Harvard University Press, 2015.

Qianlong Qingding daqing huidian zeli, vol. 150, 13–14. (乾隆 《欽定大清會典則例》卷 150，《都察院六·霸占》第 13–14 頁。)

Qingshigao 清史稿 [Draft of Qing History].

Qingshilu 清實錄 [Veritable records of the Qing dynasty].

Rawski, Evelyn. *The Last Emperors: A Social History of Qing Imperial Institutions*. Berkeley: University of California Press, 1998.

Report of Events in Peking from the 20th June to the 14th August, 1900, *Further Correspondence Respecting Events at Peking* [In continuation of “China No. 4 (1900)”]. Great Britain, Foreign Office Correspondence 4 (1900–01), June 21, p. 3.

Reynolds, Douglas. *China, 1898–1912: The Xinzheng Revolution and Japan*. Cambridge, MA: Harvard University Press, 1993.

Rogaski, Ruth. *Hygienic Modernity: Meanings of Health and Disease in Treaty-Port China*. Berkeley: University of California Press, 2004.

Rowe, William T. *Hankow: Commerce and Society in a Chinese City, 1796–1889*. Stanford, CA: Stanford University Press, 1984.

Rowe, William T. *Hankow: Conflict and Community in a Chinese City, 1796–1895*. Stanford, CA: Stanford University Press, 1989.

Sakura, Yasuo. “History of Hydrogeology in Japan.” In *History of Hydrogeology*, edited by Nicholas Howden and John Mather, 136. CRC Press, 2013.

Schoppa, Keith. *Xiang Lake: Nine Centuries of Chinese Life*. New Haven, CT: Yale University Press, 1989.

Shen, Bang 沈榜. *Wanshu zaji* 宛署雜記 [Miscellaneous notes about Wanping county]. Beijing: Beijing guji chubanshe, 1980.

Shi, Minzheng. “Beijing Transformations: Urban Infrastructure, Public Works, and Social Change in the Chinese Capital, 1900–1928.” PhD diss., Columbia University, 1993.

Shi, Minzheng. “From Imperial Gardens to Public Parks: The Transformation of Urban Space in Early Twentieth-Century Beijing.” *Modern China* 24, no. 3 (July 1998): 219–54.

Shi, Xuan. *Jiujing yishi.* (史玄：《旧京遗事》北京：北京古籍出版社，1986，第 4 页。)

Skinner, G. William “Introduction: Urban Social Structure in Ch’ing China.” In *The City in Late Imperial China,* edited by G. William Skinner, 3–5. Stanford, CA: Stanford University Press, 1977.

Smith, Carl. *City Water, City Life: Water and the Infrastructure of Ideas in Urbanizing Philadelphia, Boston, and Chicago.* Chicago University of Chicago Press, 2013.

Smith, S. A. *Like Cattle and Horses: Nationalism and Labor in Shanghai, 1895–1927.* Durham, NC: Duke University Press, 2002.

Song, Yun 松筠. *Baier laoren yu* 百二老人語 [Quotations of Bai’er old men] 首都图书馆古籍部藏.

Steinberg, Theodore. *Nature Incorporated: Industrialization and the Waters of New England.* Amherst: University of Massachusetts Press, 1991.

Strand, David. *Rickshaw Beijing: City People and Politics in the 1920s.* Berkeley: University of California Press, 1989.

Takamura, Hiroki. “The Techniques and Tools of the Kazusabori Well-boring.” *Journal of Earth Environment Research* 6 (2004): 37–49.

Tan, Qian 談遷. *Beiyoulu* 北游錄 [The record of north travel]. Beijing: Zhonghua Shuju, 1960.

Tebeau, Mark. *Eating Smoke: Fire in Urban America, 1800–1950.* Baltimore, MD: Johns Hopkins University Press, 2012.

Tiangong kaiwu (天工開物, 1637).

Timkowski, George. *Travels of the Russian Mission, through Mongolia to Peking in the Years 1820–1821,* vol. 2. London, 1827.

Unidentified author. *The Trivial Record of Beijing’s History* (出版者不詳，燕京今古瑣聞錄，民國).

Vitale, B. G. *Pekinese Rhymes: First Collected and Edited with Notes and Translation.* Peking: Pei-T’ang Press, 1896.

Wang, Di. *Street Culture in Chengdu: Public Space, Urban Commoners, and Local Politics, 1870–1930.* Stanford, CA: Stanford University Press, 2003.

Wang, Di. *The Teahouse: Small Business, Everyday Culture, and Public Politics in Chengdu, 1900–1950.* Stanford, CA: Stanford University Press, 2008.

Wang, Fu. *Wangtaichang ji,* vol. 2. Beijing: Yiwen yinshuguan. (王符撰：《王太常集》北京：藝文印書館，卷 2 遺事。)

Wang, Xitong 王錫彤. *Yizhai zishu* 抑齋自述 [Self-Recount report in Yi Hall]. Kaifeng: Henan daxue chubanshe, 2001.

Wang, Yun 王惲. *Qiujian ji*, vol. 36, Xinjing ji. (王惲：《秋澗集》卷 36，《新井記》。)

Wanshou shengdian tu. Beijing: Xueyuan chubanshe, 2001. (《萬壽盛典圖》 1717, 北京: 學苑出版社, 2001 年。) [Drawings of the sixtieth birthday procession of the Kangxi Emperor].

Weber, Heike and Stefan Krebs, eds. *The Persistence of Technology: From Maintenance and Repair to Reuse and Disposal* (transcript Verlag, 2021), 9–25.

Will, Pierre-Etienne. "State Intervention in the Administration of a Hydraulic Infrastructure: The Example of Hubei Province in Late Imperial Times." In *The Scope of State Power in China,* edited by Stuart Schram. London: SOAS, 1985.

Will, Pierre-Etienne. "Clear Waters versus Muddy Waters: The Zheng-Bai Irrigation System of Shaanxi Province in the Late Imperial Period." In *Sediments of Time: Environment and Society in Chinese History,* edited by Mark Elvin and Ts'ui-jun Liu, 283–343. Cambridge: Cambridge University Press, 1998.

Willis, Bailey. *Research in China*, vol. 1, pt. 1. Washington, DC: Carnegie Institution of Washington, 1907.

Wittfogel, Karl A. *Oriental Despotism: A Comparative Study of Total Power.* New Haven, CT: Yale University Press, 1967.

Wu, Boxiao. "Wuhuaguo-Wo he sanwen." *Wenxue pinglun*, no. 5 (1981): 70–75. （吳伯簫：無花果-我和散文，文學評論，1981 年第 5 期，70–75。）

Wu, Tingxie 吳廷燮, ed. *Beijing shi zhigao* 北京市志稿 [Manuscripts of the Beijing Gazetteer], vols. 1–15. Beijing: Yanshan chubanshe, 1998.

Wu, Zhengge 吳正格, ed. *Manqing shisu yu qinggong yushan* 滿清食俗与清宮禦膳 [Manchu's diet and imperial dietary in palace]. Shenyang: Liaoning kexue jishu chubanshe, 1988.

Wu, Zhenyu 吳振域. *Yangjizhai conglu*養吉齋叢錄 [Collections of Yangji Hall]. Beijing: Zhonghua shuju, 2005.

Xiong, Yuanbao. "Shindai Pekin no suidoji" [Water distribution territories in Qing and Republican Beijing]. In *Social history of Qing Huizhou,* edited by Xiong Yuebao, 291–327. Tokyo, 2003).

Xiong, Yuangan 熊元幹, ed. *Beijing defang shenpanting facaohui pandu huibian* 京師地方審判廳法曹會判牘匯編 [Compilations of Files in Beijing municipal court]. Beijing: Jingshi defang shenpanting, 1914.

Xiong, Yuangan 熊元幹, ed. "Sun Shufa konggao Liang Zongxin yin shujing jiuge yian panic." In *Jingshi defang shenpanting facaohui pandu huibian.* Beijing, 1914.

Xu, Duanqing 許端慶. "Beiping zhi gonggong weisheng yipie" 北平之公共衛生一瞥 [A glance at public health in Beijing]. *Tongji yixue jikan* 同濟醫學季刊 [TUNG-CHI ACTA MEDICA] 4, no. 1 (March 1934): 70–80.

Xu, Yamin. "Wicked Citizens and Social Origins of China's Modern Authoritarian State: Civil Strife and Political Control in Republican Beiping, 1928–1937." PhD diss., UC Berkeley, 2002.

Xu, Yamin. "Policing Civility on the Streets: Encounters with Littergus, 'Nightsoil Lords,' and Street Corner Urinators in Republican Beijing." *Twentieth-Century China* 30, no. 2 (2005): 28–71.

Xu, Yamin. "Urban Communities, State, Spatial Order, and Modernity: Studies of Imperial and Republican Beijing in Perspective." *China Review International* 15, no. 1 (2008): 1–38.

Yang, Daqing. *Technology of Empire: Telecommunications and Japanese Expansion in Asia, 1883–1945.* Cambridge, MA: Harvard University Press, 2011.

Yang, Nianqun 楊念群. *Zaizao Bingren: Zhong Xi Yi Chongtu Xia De Kongjian Zhengzhi, 1832–1985* 製造病人 (Remaking "Patients": Politics of Space in the Conflicts between Traditional Chinese Medicine and Western Medicine, 1832–1985). Beijing: Remin University Press, 2006.

Yanjing zaji. Beijing: Beijing Guji Chubanshe, 1986. (闕名：燕京雜記 北京：北京古籍出版社，1986 年。)

Yao, Xunyuan 姚尋源. "Cangying jingshui lengying yu beepingren changwei zhi guanxi" 蒼蠅井水冷飲与北平人腸胃癥之關繫 [Flies, well water, cold beverage and Gastrointestinal disorder in Beiping]. *Yixue zhoukanji*, 醫學週刊集 [*Weekly Journal of Medicine*] 3 (1930).

Yu, Xiezhong 余協中. "Beiping de gonggong weisheng" 北平的公共衛生 [Beiping's public health]. *Shehui xuejie* 社會學界 [*Sociological World*], no. 3 (September 1929): 63–82.

Zelin, Madeleine. *The Merchants of Zigong: Industrial Entrepreneurship in Early Modern China.* New York: Columbia University Press, 2005.

Zhang, Cixi 張次溪. *Yanjing fanggu lu* 燕京訪古錄 [The records of visiting ancient place interests in Beijing]. Beijing: Zhonghua yinshuju, 1934.

Zhang, Daqing. *Daqing xinfaling* 大清新法令 [New laws in Qing Dynasty 1907]. *Zhongguo jindai jibing shehuishi, 1912–1937.* Jinan: Shandong jiaoyu chubanshe, 2006, 86. (《大清新刑律》1907, 張大慶：《中國近代疾病社會史，1912–1937》濟南：山東教育出版社， 2006 年，第 86 頁。)

Zhang, Dechang 張德昌. *Qingji yige jingguan de shenghuo*, 清季一個京官的生活 [The Life of an official in Beijing in late Qing]. Hong Kong: Chinese University of Hong Kong Press, 1970.

Zhang, Jian 張謇. *Zhang Jian Quanji* 張謇全集 [*Complete works of Zhang Jian*]. 南京：江蘇古籍出版社，1994 年.

Zhang, Ling. *The River, the Plain, and the State: An Environmental Drama in Northern Song China, 1048–1128.* Cambridge: Cambridge University Press, 2016.

Zhang, Wu. *Zhengli Beijingshi jihuashu*. Beiping, 1928, 27–28. (張武著：整理北京市計劃書 ，1928)

Zhang, Xiande. "Shuijing de lishi ji Beijing chutu shuijing gaishu." In *The General History of Wells in Beijing,* edited by Su Tianjun, 274–75. Beijing: Beijing chubanshe, 2000. (張先得《水井的歷史及北京出土水井概述》，蘇天鈞編: 《北京考古集成》，北京：北京出版社，2000 年，第1卷， 第 274–275 頁。)

Zhang, Xiaolin 張小林. *Qingdai Beijing chengqu fangqi yanjiu* 清代北京城區房契研究. Beijing: Zhongguo shehui kexue Chubanshe, 2000.

Zhao, Jianai. "Beiping tebieshi gongyongju 1929 niandu shizheng dagang." *Beijing dangan shiliao*, no. 1 (1992): 7–9. (趙家鼐選編：北平特別市公用局1929年度施政大綱，北京檔案史料 1992 年第 1 期, 7–9。)

Zhongguo diyi lishi danganguan, ed. *Manwen laodang* 滿文老檔 [Manchurian old archive]. Beijing: Zhonghua shuju, 1990.

Zhongguo diyi lishi danganguan, ed. "Qingmo Beijing kuochong zilaishui zhuanguan shiliao." *Lishi dangan*, no. 2 (1992): 64. (中國第一歷史檔案館，清末北京擴充自來水專管史料，歷史檔案 1992 年第 2 期，第 64 頁。)

Zhongguo renmin zhengzhi xieshang huiyi quanguo weiyuanhui ziliao yanjiu weiyuanhui, ed. *Wanqing gongting shenghuo jianwen* 晚清宮廷生活見聞 [Witnesses of court lives in the late Qing]. Beijing: wenshi ziliao chubanshe, 1982.

Zhou, Shuzhen 周淑娟. *Zhou Zhian Xiansheng biezhuang*周學熙先生別傳 [Biography of Zhou Zhian]. Taipei: Weihai chubanshe.

Zhu, Yixin 朱一新. *Jingshi fangxiang zhigao* 京師坊巷志稿 [*The gazette of lanes in Beijing*]. Beijing: Beijing guji chubanshe, 1982.

Index

Printed in the United States
by Baker & Taylor Publisher Services